A Demo A Day

A Year of Physics Demonstrations

Borislaw Bilash II

David P. Maiullo

Published by:

FLINN SCIENTIFIC, INC.
P.O. Box 219
Batavia, IL 60510
(800) 452-1261

ISBN 978-1-933709-16-1

Dedication

To "Christine," our muse . . .

As we embarked on this project, we gave some hard thought about who would be our readers, what they would most want and need in such a book and how best to convey that information. In doing so, we envisioned the physics teacher we'd be writing for and kept her in mind throughout the writing of the book. This teacher is a first-year novice, full of enthusiasm and seeking to become the best physics teacher possible. While ambitious, she has limited experience in building equipment and performing demonstrations. To make her as fully real to us as possible, we gave her a name—Christine. As we wrote each demo, we often asked ourselves: What would Christine need to know? Can Christine build this? And other questions such as: Does Christine know what a furring strip is? This book is dedicated to Christine and all the physics teachers who strive to be their very best.

Table of Contents

Chapter 20 — Interference and Diffraction

Chapter 21 — Atomic and Nuclear Physics

Dr. Warren W. Hein, Executive Officer

American Association of Physics Teachers

The use of demonstrations in physics classrooms to increase student interest and contribute to the understanding of the physics principles under discussion has a long history. The story of Oersted's discovery linking electric current to magnetism is said to have come about as he was doing a demonstration before a group of students. Whether this is true or not, a good demonstration presented in a sound pedagogical manner can be very effective and lead to increased qualitative as well as quantitative understanding.

What is good pedagogy when it comes to demonstrations? Good teaching instructions don't begin and end at the "gee whiz" moment of a demonstration. (A former student once told me the only thing she remembered about my class was when I submerged a hot dog, banana and rubber hose in liquid nitrogen and shattered them with a hammer.) Good pedagogy seeks to explain the theory or theories behind the demonstration and offer not only ways to help impart the lesson to students but also how to get students to think through for themselves what's being demonstrated so that they can more fully understand and grasp the principles.

We know that doing a demonstration should be more than just "show," but requires engagement of the students in the activity. This can be done by having the individual students predict ahead of time what is going to happen, discussing it with their peers and maybe even voting on the expected result before the demonstration is actually "performed." It has been shown that significant gains in student learning can be achieved by getting students engaged in the demonstration in this manner. The Physics classroom is not the place for magic shows, but instead should be a theater where students see the principles of science become connected and evolve into a working wonder we call our universe.

There have been a number of books written that provide compendia of demonstrations that can be used in the pre-college and college/university classroom. This new book, "A Demonstration A Day," by Maiullo and Bilash provides a unique resource for the pre-college teacher with a set of well-chosen demonstrations and guided inquiries that can be used throughout the entire school year. Among the things that make this book different and outstanding is that the demonstrations can be easily constructed from readily available and low-cost or no-cost materials. But more important, the theory or theories behind each demo are clearly explained and the authors supplement their descriptions with questions to ask students to help them figure out for themselves the principles at work, making each demonstration more than just a "gee whiz" sideshow but an engaging learning experience, meant to get students excited about learning and passionate about science. What's more, their instructions are so detailed that a teacher could assign a student to build many of the demos, creating an additional learning opportunity for the student and also cutting teacher prep time.

Demonstrations in the physics classroom, whether it is a pre college level or college level course, have proven to be valuable means of engaging students and initiating discussion. A good and meaningful demonstration, just as a picture, can be worth 1,000 words. When students realize the science behind the mystery they feel empowered and begin to see the world around them in a different light.

Maiullo and Bilash are to be congratulated for sharing their many years of experience in the area of classroom demonstrations to make this resource available to pre-college physics and physical science teachers.

Warren Hein

About the Authors

Borislaw Bilash II, B.Sc, EdM.

Borislaw has taught high school physical science, chemistry and physics in New Jersey since 1986. Currently he teaches physics at Pascack Valley High School in Hillsdale, NJ. Born and raised in Winnipeg, Manitoba, Canada, Borislaw received his B.Sc. (Physics) from the University of Manitoba and his Ed.M (Science Education) from Rutgers University. Over the years he has taught freshman chemistry at the New Jersey Institute of Technology, and worked with inner-city students in various enrichment programs including INROADS, Upward Bound and the CHIME programs at local universities. He is the lead instructor at the Physics and Chemistry Institutes for High School Teachers at the Liberty Science Center in Jersey City, NJ. He has developed curriculum materials for the Manitoba Department of Education, Silver, Burdett, Ginn, Inc. and Prentice Hall, Inc. As principal author, Borislaw has written four books of the A Demo A Day™ series—demonstration handbooks for teachers of chemistry, biology and physical science. He has coordinated the Demo Dens at the NJ Science Convention since 1992 and is a regular presenter at state and national conventions often presenting on the topic of demonstrating science and using the constructivist and inquiry methods of teaching science. Borislaw is a Tandy Scholar and has been recognized by numerous organizations including the NJ Science Teacher's Association, NJ Institute of Chemists, the American Chemical Society and USA Today's All USA Teacher Team. Borislaw is a NJ Governor Fellow of Math and Science.

David P. Maiullo, B.S.

David has been working as a Physics Support Specialist at Rutgers University supervising the Department of Physics and Astronomy lecture demonstration facility since 1986. David has been active in the New Jersey Section of the American Association of Physics Teachers (NJAAPT), coordinating workshops and demonstration shows and serving on its Executive Board since 1990. David is recognized as a demonstrator extraordinaire and for his work in advancing the craft of physics lecture demonstrations as a member of the Apparatus Committee of the American Association of Physics Teachers (AAPT), and as a leader in the Physics Instructional Resource Association (PIRA). He has served as the chair of the Apparatus Committee and president of PIRA. David is a regular presenter at state and national conventions—teaching teachers how to develop, construct, and present lecture demonstrations for all levels of physics education. He frequently conducts public physics demonstration shows at street fairs, bars, libraries and schools in and around the metropolitan New York City area—having been featured on the *New York Times* Web site. David has developed a video series of physics demonstrations for Wiley & Company that is distributed on DVD and accessible through Wiley's online learning. Rutgers University has recognized David's work with the Ernest E. McMahon Award for Public Outreach and the President's Excellence in Service Award. In 2006 David received the Lifetime Service Award from the New Jersey Section of the American Association of Physics Teachers.

<h1 align="right">Acknowledgments</h1>

Few of the ideas in this book are original. Many of the ideas originated on the pages of the Physics Teacher, the American Journal of Physics, in addition to the iconic physics demonstration books: *Demonstration Experiments in Physics* (1938) by Richard Sutton and *The Dick and Rae Demonstration Notebook* (1993) by Dick Minnix and Rae Carpenter. These resources are invaluable.

We are greatly indebted to our colleague Professor Mark Croft of The Physics Department at Rutgers University for his encouragement and suggestions as our sounding board and, foremost, for his friendship. We thank our fellow teachers of the New Jersey Section of the American Association of Physics Teachers and our colleagues from the New Jersey Science Teachers Association for their inspiration and support of our demonstration workshops and for their constructive feedback. Many of the ideas contained in this book were fostered through the interactions with our colleagues at Physics Instruction Resource Association (PIRA) and on the pages of the Internet Listserv known as "Tap-L." We are grateful to Dr. Richard (Dick) Berg, Director of the Physics Lecture–Demonstration Facility at the Department of Physics, University of Maryland, for sharing his ideas and experience over the years.

John K. Koob, a colleague, a mentor and a good friend. His spirit is found throughout this book.

Bill Koenig, physics teacher at Pascack Valley High School has been instrumental as our test driver and critic, and helped vet many of these ideas. Bill, thank you for lending us your ear (and your tools), for challenging us with good questions, and for your tolerance of the baggage that accompanies a creative journey.

Our gratitude is extended to Irene G. Cesa, Director of Technical Services at Flinn Scientific, for her expertise, support and encouragement and for serving as the heart of production. We appreciate the dedicated efforts of Shirley Wehner in formatting the book and thank her for creating the artwork found throughout the book. Thanks also to Alyssa Bevans for ensuring that our writing conveys our thoughts as they were intended. Foremost, we thank Karen Kihlstrom, Scientific Support Specialist, our editor, for the essential and critical review of the physics concepts presented in this book, for leaving no stone (nor principle of physics) unturned and for testing many of the demonstrations in her lab. Her expertise and contributions have been very helpful.

There were many weekends spent in the Catskill Mountains writing within sight of the Schoharie Creek. Nature is so precious and encouraging.

Finally, we are thankful to our wives, Natalia and Laura, for reviewing our manuscript and for all of the love and support they provided while we worked to perfect the demonstrations contained in this book. We couldn't have done this without you.

General Safety Guidelines for Physics Demonstrations

1. Safety is paramount. Always practice the demonstration before performing it in front of your audience to verify that neither the audience, nor the performer, is at risk of injury.

2. Safety goggles should be worn whenever there is risk of eye injury—including, but not limited to, working with chemicals, including liquid nitrogen, heating objects or liquids, working with power tools, and demonstrations using projectiles or springs, etc.

3. Wear hearing protection, and instruct students or other audience members to cover their ears whenever loud noises will be emitted during a demonstration.

4. Protect extremities, including hands and feet, when lifting, dropping or moving objects.

5. Use tongs or insulated gloves whenever moving hot objects such as hot plates, burners, heated beakers, etc.

6. Have a fire extinguisher available whenever performing demonstrations involving flammable substances, open flames or sparks.

7. Follow manufacturers' directions whenever working with power tools.

8. Whenever evacuating a container, such as a bell jar or penny-feather tube, always verify by close visual inspection that the container is free of cracks or chips. Limit vacuum pressures to a safe minimum.

9. Always ensure that electrical devices are properly grounded.

10. Capacitors store electric charge and may be dangerous, even when disconnected from an electric circuit. Discharge capacitors that are to be stored or may be unattended.

11. Whenever working with electricity, ensure that the current is shut off prior to making adjustments to the equipment or circuit.

12. Refrain from using alternating current to power exposed circuits in demonstrations.

13. Limit direct current circuit demonstrations to short pulses not exceeding 20 VDC and 10 amps.

14. Keep high strength magnets, such as neodymium magnets or electromagnets, away from magnetic media and individuals with pacemakers.

15. Avoid touching radioactive substances. Use tongs to handle radioactive samples. Radioactive samples should be selected such that they emit the minimum amount of radiation—just enough to register an observation (1 μCu maximum).

16. Do not look directly into a laser beam. Limit the power of demonstration lasers to <5 mW.

Best Practices for Demonstrating Science

Demonstrations can be used to introduce a concept, to serve as an example or to verify the understanding of a concept. Regardless of your approach it is recommended that the demonstrator:

1. **Practice the demonstration** to ensure that it is safe for both the audience and the demonstrator AND to ensure that the demonstration works consistently.

2. **Prepare the demonstration area** ensuring that it is well lit and that the audience can see and hear all aspects of the demonstration. If necessary, use a video camera and projector to zoom onto hard-to-see apparatus—the student sitting in the nosebleed seats should enjoy the same experience as those in the splash zone. A microphone and sound system can be used to enhance the audio of the presenter and the demonstration.

3. **Have all safety equipment within reach** of the demonstration area, including equipment such as a fire extinguisher, and verify that your electrical outlets and extension cords have GFI protection. If an extension cord must be used, be sure that it does not pose a risk of tripping over it or getting caught by either your audience or by you while performing the demonstration.

4. **Demonstration areas should be free of clutter** and demonstration equipment should be laid out in an organized fashion to facilitate an efficient performance of the demonstration. A cart may be used to organize the equipment and to move the equipment from place to place.

5. **Address a specific learning objective** so that the audience is left educated—not simply thrilled. Your goal is to teach!

6. **Involve the audience** as much as possible by having them respond to questions of inquiry so that the experience is not passive. Take advantage of the natural curiosity of your audience. A Chinese proverb says: Tell me—I forget. Show me—I remember. Involve me—I understand. Make this your credo!

7. **Refrain from "Teaching Gray Using an Elephant."** Students are apt to miss the point if you were to bring an elephant into your classroom to teach the concept of gray. A poorly chosen prop can overpower what you are trying to accomplish. The demonstration selected should appropriately address the learning objective.

8. **Show your enthusiasm** when performing the demonstration in order to draw students' interest. Make your enthusiasm contagious. If you are excited, the audience will be excited!

9. **Learn something new by reading about demonstrations** found in professional literature including *The Physics Teacher* or on Web sites hosted by university physics departments and Listserves.

10. **Network with others to learn** about demonstrations and to share ideas. Become an active member of professional groups including the American Association of Physics Teachers (AAPT), the National Science Teachers Association (NSTA) and your state or provincial science teachers association. Each of these groups sponsor workshops, talks and sharing sessions on demonstrations. Don't be shy, and get involved. Hone your craft and demonstration style by presenting your ideas to other teachers. There is no better way to learn than by doing.

Chapter 1

Introducing Physics and Measurement

How High?

The height of the flight of a rocket is determined from a distance using trigonometric techniques.

Application | Trigonometry • Measurement techniques • Nautical instruments

Theory | The height of an object, such as a tree, building or low flying object, may be determined using a technique that employs basic trigonometric functions. The technique involves creating an imaginary right-angled triangle in which the observer stands at a distance "x" from the point that lies directly underneath the object. If the object is a tree or building, x represents the distance between the observer and the base of the tree or building. The line of sight between the observer and the height of the object forms an angle θ with respect to line "x." The line of sight represents the hypotenuse of the triangle while "x" represents the base of the triangle (see Figure 1). The height of the object, H may be determined by measuring θ with a sextant (altitude finder) and "x" with a meter stick or tape measure. The line of sight is determined by looking through the tube or sites of the sextant. The height is then calculated using $H = x/\arcsin\theta$. This technique is reasonably accurate for low altitude objects.

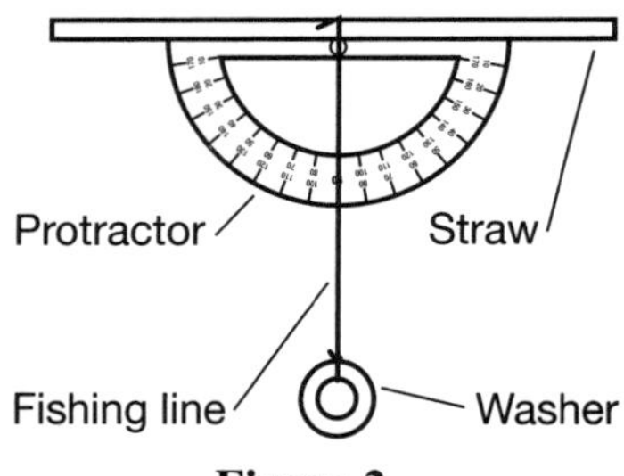

Figure 1.

Materials | Calculators, one for each pair of students

Field, large, preferably a marked football field

Rocket, air-, water- or chemical-powered

Sextant (altitude finder)

Tape measure or meter stick

Note: A simple sextant can be made using a protractor, straw, fishing line, washer and hot glue (see Figure 2).

Figure 2.

Safety Precautions | Although this demonstration does not pose any obvious risks, always follow laboratory safety rules.

Demonstration | In the classroom, review with students how to determine the height of an object using a ruler and the angle of sight. Instruct students how to use a sextant to measure the angle of sight.

Determine the height of the football goal post (or flag pole).

Place the rocket at the end of a field and have students stand 50–100 meters away. Have students measure the exact distance between their position and the rocket. If a marked football field is used, students can determine the distance by converting the yard lines to meters (1 yd = 0.92 m). Instruct students to measure the angle of sight for the maximum height achieved by the rocket. Have students calculate the height of the rocket using trigonometry.

The diameter of the Sun is measured from a distance using simple trigonometric techniques.

Application | Trigonometry • Measurement techniques • Nautical instruments

Theory | Making measurements of large distances using basic mathematical techniques dates back thousands of years. For example, Eratosthenes (276–195 BC) measured the radius of the Earth and Ptolemy (100–170 AD) approximated the distance between the Earth and Moon. In this demonstration, students will cast an image of the Sun. The image diameter and the image distance can be compared to the Sun's diameter and distance using the technique of Similar Triangles. In summary,

$$\frac{\text{Image distance}}{\text{Image diameter}} = \frac{\text{Sun distance}}{\text{Sun diameter}}$$

The best time to perform this demonstration is between the hours of 10 am and 2 pm (as close to noon as possible). It can also be done earlier or later, however, with adequate results.

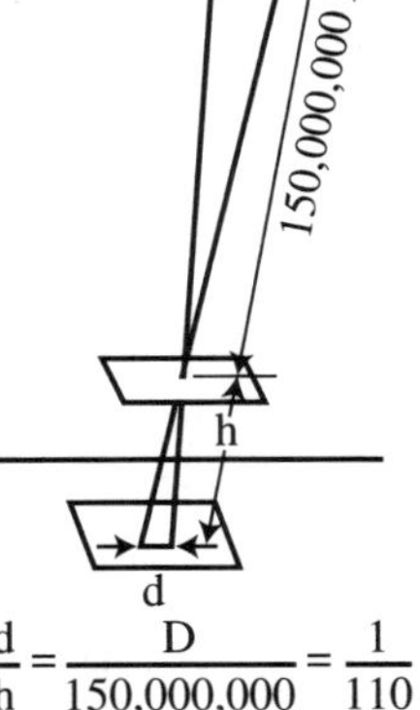

$$\frac{d}{h} = \frac{D}{150,000,000} = \frac{1}{110}$$

Materials | Index cards, 2

Meter stick

Pin

Optional: Graph paper with 1–2 mm increments and tape

Safety Precautions | This demonstration is not considered hazardous, however, remind students not to look directly at the Sun.

Preparation | Make a pinhole in the center of one index card. As an option, tape a 5 × 5-cm piece of 1–2 mm increment graph paper on the second index card.

Demonstration | In the classroom, review the techniques of measuring objects from a distance.

Outside in view of the Sun, instruct students to lay the index card with the graph paper on the ground. This index card will serve as a screen for the demonstration. Hold the card with the pinhole above the screen. An image of the Sun should be visible. Use a meter stick to measure the diameter of this image and the distance between the two cards.

Have students calculate the diameter of the Sun, given that the average distance between the Earth and the Sun is 150,000,000 km.

Reference | Riordan, Robin. *Measuring the Size of the Sun;* Prince George Centre. Royal Astronomical Society of Canada; 2004.

Freeze!

Demonstrate how a stroboscope can be used to "stop" rapid motion.

Applications | Measurement techniques • Light • Sound

Theory | A stroboscope is a light that is similar to a photographic flash unit but that can produce flashes at a rate as high as 10,000 flashes per second. When used in a darkened room, the flash of a stroboscope can be used to "freeze" the movement of an object. Stroboscopes are commonly used at social dances to create an interrupted motion effect as people move across the dance floor. The same technique may be employed to study the rate of motion of an object. The stroboscope reveals the position of a moving object as it moves through the dark. The speed of an object moving at a constant velocity may be determined by measuring the change in position between revealing flashes and dividing it by the rate (i.e.: time) of the flash unit. The changing position of the object may be preserved by taking a photograph with the film being exposed during the entire movement resulting in a multiple exposure effect. In this demonstration, a stroboscope is used to "freeze" the movement of a fan in order to read a message written on its blades. Harold "Doc" Edgerton (1903–1990) of the Massachusetts Institute of Technology developed techniques, called ultra-high speed and stop-action photography, that utilized the stroboscopic effect to analyze moving objects. The *Balloon and the Bullet* and the *Milk Drop Coronet* are two of his most widely seen photographs.

Students should be reminded of the appearance of car or wagon wheels in a movie, where they appear to rotate backwards. In the case of a movie, the rate of the still images captured produces a stroboscopic effect that makes the rotating objects appear to be moving in the reverse direction. The perception of backward rotation has also been seen under continuous illumination such as when viewing the seemingly backward movement of forward moving Mag-wheels of a car or propeller of a plane as the plane's engine is shut off or turned on. It is believed that human visual perception takes a series of still frames, the same way a strobe illuminates an object in a series of still frames.

Materials | Fan, personal 2-battery cooling (available in summer months) or regular fan

Marker

Ring stand

Stroboscope, electronic variable or hand-held

Tuning fork (optional)

Utility clamp

| **Safety Precautions** | Although this demonstration does not pose any obvious risks, always follow laboratory safety rules. Speak to your school nurse to verify if any of your students have a history of seizures since the flashing of a stroboscope has been known to initiate a seizure in some people. The stroboscope will make the fan blades only appear not moving or moving at slower speeds. For this reason remind students never to attempt to touch or insert objects into the moving blades. |

Preparation

Use a marker to write a phrase, such as "Physics is Phun!" on the blades of the fan. Secure the fan to a utility clamp on a ring stand.

Demonstration

With the fan initially facing away from students, start the fan motor. Now face the fan towards the audience. The phrase written on the fan blades will not be readable. If using the electronic strobe, darken the room slightly. Adjust the frequency of the strobe until the propeller appears to stop. Once the blades are "frozen" the message may be read. By changing this frequency, you could make the propeller appear to move in either direction.

The stroboscopic technique can further be illustrated by slowing or stopping the motion of a vibrating tuning fork in a dimly lit room.

Reference

Based on an exhibit at the Boston Museum of Science, May 2005.

Purves, D., Paydarfar J., Andrews T. "The Wagon Wheel Illusion in Movies and Reality"; Proceedings of the National Academy of Sciences USA. 93 (8), pp 3693–7.

Lazy Walls

A vibrometer is used to magnify the minute compression of a wall.

Application | Measurement techniques • Seismographs • Earthquakes

Theory | Making measurements often requires the use of techniques or tools that will magnify or amplify the measurements in order to observe them. A vibrometer is a device that allows us to make an indirect measurement of a vibration. A seismometer is an example of a vibrometer. In this demonstration a vibrometer is constructed to amplify the movement of a wall due to a person applying pressure to it with his/her hand. The technique employs a rod with one end touching the wall and the opposite end resting on a pin that is attached to a mirror. When the wall is compressed, the pin will turn, causing the mirror to rotate. A laser beam focused on the mirror will reflect a beam that will deflect dramatically as the wall is compressed.

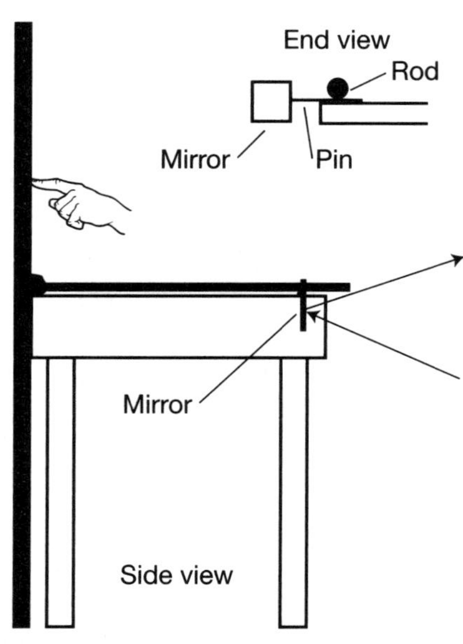

Materials

Adhesive, construction (or epoxy)	Ring stand
Clay, modeling, a lump the size of a quarter	Table
Iron rod, 1-m (from a ring stand)	Tape, masking
Laser pointer	Utility clamp
Mirror	Wall
Pin, straight	

Safety Precautions | Do not direct a laser beam into eyes. A laser produces an intense beam of light that may cause permanent damage if directed toward an eye.

Preparation | Glue a straight pin in the middle of the back of a small square mirror. Lay a ring stand rod on the edge of a table that is situated against a wall. Adhere one end of the rod to the wall using a lump of modeling clay. Slip the pin, which is attached to the mirror, between the rod and the table, so the mirror hangs over the edge of the table. The mirror should be situated as far away from the wall as possible.

Demonstration   Ask students if it is possible for a human to move a cement wall. Expect many students to be skeptical that it can be done. Direct students' attention to the vibrometer you have constructed. Aim a laser at the mirror such that the reflection of the beam lands on a wall. Place a piece of masking tape on this spot, which will serve as a reference. If using a laser pointer, secure the laser with a utility clamp attached to a ring stand. Use your finger to move the wall by pressing on the wall about a meter from where the rod is attached with modeling clay.

Reference   Special thanks to Harry Rheam of Eastern Senior High School in Atco, NJ for sharing this idea.

Tea in China

Application | Order of magnitude • Estimating • Fermi problems

Theory | A Fermi Problem or Back-of-the-Envelope Estimate is a problem in which a reasonable estimate is possible using the technique of order-of-magnitude approximation. Named for Enrico Fermi (1901–1954), the problem-solving technique involves making justified guesses of quantities that may be impossible to compute due to limits in information. The final estimate would be within an order of ten of the actual answer. Fermi used the technique often to determine quantitative answers to questions that he had no other means of answering. In one instance, Fermi was able to estimate the strength of the atomic bomb detonated at the Trinity site based on pieces of paper he dropped during the blast. In addition to those questions of serious consequence, Fermi was known to enjoy toying with and challenging his colleagues with questions such as 'How many piano tuners are there in Chicago?'. Fermi even tackled the cliché "I wouldn't do it for all the tea in China" by estimating how much tea is in China.

The approach to a Fermi problem requires us to think about the assumptions we make and to make those assumptions as realistic as possible, as well as make good estimates and put those estimates together into a logical sequence of calculations to arrive at the final answer. Practicing Fermi Problems helps develop analytical skills and out-of-the-box thinking that lead to insight.

Answering Fermi questions is common in physics Olympic competitions on a high school level.

Materials | Large jar of rice

Safety Precautions | Although this demonstration does not pose any obvious risks, always follow laboratory safety rules.

Demonstration | Discuss the frequent need for scientists to make estimations and the technique of identifying order of magnitude to accomplish the task.

Ask students, how much tea is there in China? Show students how to make a reliable estimation:

Assumptions

- There are 1 billion or 10^9 people in China.

- Each person in China drinks, on average, 0.5 L of tea per day. It takes maybe 10 g or 10^{-2} kg of tea leaves to make that.

- There is a three-month or 90-day supply of tea on hand in China at any given time. Heck, call it a 10^2 day supply.

Calculation

$(10^9 \text{ people})(10^{-2} \text{ kg/person/day})(10^2 \text{ days}) \ = \ 10^9 \text{ kg of tea}$

That is a *billion* kg of tea!

Now show students a large jar of rice and have them determine the number of grains in the jar.

Finally, try some of the following Fermi Problems:

1. How many frames are in a Walt Disney animated movie such as *Tarzan*?

2. How many piano tuners would you expect to find in the local telephone directory?

3. If all the oxygen atoms breathed by Enrico Fermi over his lifetime are now distributed uniformly through the atmosphere, how many of these atoms do you breathe in with each breath?

4. If you could get a penny for each time someone said "Ouch!" in the United States, how long would it take you to become a billionaire?

5. Pick a nearby tree and estimate the number of leaves on the tree.

How'd You Do That?

Practice using observations to make hypotheses.

Applications | Scientific method • Hypothesis • Observations

Theory | After gathering all available information on a problem, scientists formulate possible solutions to problems called hypotheses. The process requires careful analysis of all available data and should never consist of an empty guess. A hypothesis need not be found correct. It simply provides direction for the scientist during the course of an experiment. Provide students with a clear understanding that they should not be concerned with trying to "prove" a hypothesis correct. Instead, the hypothesis should be used to guide experimental design. Simply put, an experiment is conducted to verify the validity of a hypothesis. If the hypothesis correctly predicts observations, the scientist has gained knowledge. If the hypothesis shown to be incorrect, the scientist has still learned something which could lead to a new hypothesis and further experimentation. Sometimes learning what isn't true is just as important as learning what is true.

The apparatus used in this demonstration consists of a simple siphon surrounded by extra materials. The extra pieces of equipment (hoses, stoppers, etc.) provide other variables for students to consider in developing a hypothesis. Water poured into the funnel displaces air in the lower bottle. In turn, this air is pushed into the upper bottle and displaces the water in the upper bottle causing it to flow through the glass tube into the funnel. Once water begins to flow out of the glass tube, a siphoning action begins. The process continues until the upper bottle is emptied. To make the apparatus appear perpetual, tell students that they have three minutes to make observations and then the apparatus will be removed from sight.

After viewing the mystery machine for a few minutes, most students will agree that it appears that the water poured into the funnel appears flows through the system and back into the funnel perpetually. Their first reaction will be that "this is impossible"—but it seems to be perpetual. Most students will (incorrectly) assume that the machine will continue forever. The apparatus can be adjusted to flow for 5–10 minutes.

Materials | Mystery Apparatus:

Beaker, 500-mL	Funnel, small
Bunsen burner	Rubber stoppers, two-hole, 2
Cans or bottles, 4-L, painted with opaque paint, 2	Tubing, glass, 1 m long
File	Tubing, rubber, 50 cm long

Safety Precautions | Wear chemical splash goggles and always follow laboratory safety rules when performing demonstrations.

Preparation

Construct a mystery apparatus as in the diagram on the right.

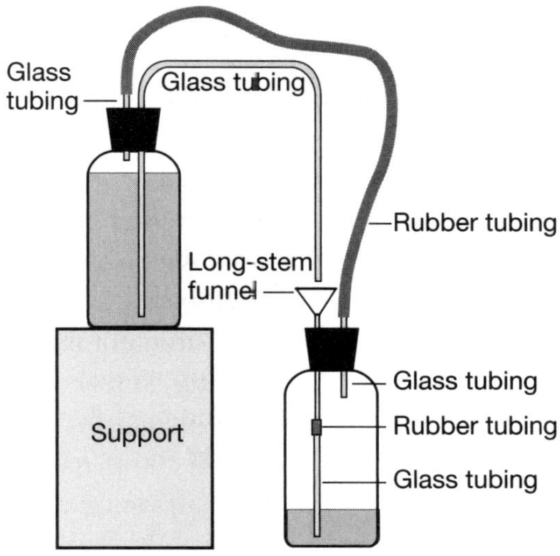

Fill the supported bottle full of water. Secure the rubber stoppers so they are airtight.

Insert a long-stem funnel into a two-hole rubber stopper. Lubricate the glass with glycerin, if necessary. Use a file to score and cut a piece of glass tubing long enough to extend from the tip of the funnel to the bottom of the 4-L bottle. Fire-polishing the ends of the tube may be desirable, but not necessary. Use a 3-cm piece of rubber tubing to connect this tube to the stem of the funnel. Score and cut two 5-cm lengths of glass tubing. Insert one piece through the second hole of the stopper funnel assembly. The tube should protrude equally from the top and bottom of the stopper. The second short glass tube short be inserted into one of the holes of the second stopper. Use a Bunsen burner or propane torch to create a U-shaped form from the remaining piece of glass tubing, as illustrated in the diagram. Insert the U-tube into the second stopper. Connect the tubes protruding from both stoppers using a 50-cm length of rubber tubing. The lengths of the glass and rubber tubing may be adjusted to match the dimensions of the bottles. Finally, fill the bottles with water and arrange them as illustrated in the diagram. Secure the rubber stoppers so they are airtight.

Demonstration

Pour about 500 mL of water into the funnel—until water begins to pour out of the open glass tube. The water will continue cycling through the system for quite a while.

Have students attempt to explain what is happening in the mystery apparatus and how it works. Have them design an experiment to test their hypothesis.

Note: Do not reveal the mystery of this apparatus. Remember, this is to serve as a model for what "real" scientists encounter. Scientists who form hypotheses regarding abstract phenomena can't ask someone if "they are right."

Disposal

None required.

Reference

Hewitt, P. *Conceptual Physics Laboratory Manual;* Addison-Wesley: Menlo Park, CA, 1987; p 1.

Reprinted from *A Demo A Day—Chemistry, Vol. 2;* Flinn Scientific: Batavia, IL, 1995 with permission.

How Close Can You Get?

Demonstrate how the concept of significant digits arises.

Application — Significant digits • Measurements • Units • Accuracy

Theory — Physical properties are often measured in standardized quantities called units of measurement. Any number that is the result of a physical measurement has a dimension associated with it. The number and the dimension are inseparable and without the dimension, the number has no meaning. An exception to this is when the measurement is based on a ratio of two quantities of identical units, which consequently cancel out resulting in a dimensionless measurement. (Examples include the coefficient of friction, which is based on a ratio of forces, and specific gravity, which compares densities.) Numbers that result from physical measurements also have an uncertainty associated with them. This uncertainty is due to the sensitivity level of the instrument used to take the measurement. Instruments with finer calibrations produce measurements with less uncertainty, but uncertainty is always present.

Materials — Assorted objects to measure

Felt pen

Meter sticks, 2

Tape, masking

Safety Precautions — Although this demonstration does not pose any obvious risks, always follow laboratory safety rules.

Preparation — Cover both sides of one meter stick with masking tape. Label one side "meters," and place a 0 at one end and a 1 at the other end. Mark the other side off in decimeters and number 0 to 10. Cover one side of the other meter stick, mark off centimeters and label 0 to 100. Leave the other side as is, marked in millimeters. As an option, a four-sided meter stick prepared as described above is commercially available.

Demonstration — Using the stick marked as one meter, measure the length of an object like a table. Choose this object so that that the measurement comes out somewhere between a whole number of meters (example—between 2 and 3 meters). Point out that a measurement recorded as a whole number would be an obvious error, but that an estimate such as 2.4 meters (or whatever it comes out to be) would be more accurate, even though the last digit is an approximation. The measurement "2.4 meters" has two significant digits—one that is undoubtedly correct and a final digit, which includes some error due to its approximation. Repeat the measurement with the decimeter scale, estimating to the nearest cm. Repeat with the centimeter scale, estimating to the nearest mm. Finally, repeat with the millimeter scale, estimating to the nearest 0.1 mm.

This exercise may be used as an introduction to significant digits. Point out to the students that sooner or later we will run out of measuring units, and therefore the accuracy of a measurement is limited.

Reference

Special thanks to John Koob (deceased), formerly of Queen of Peace High School in North Arlington, NJ, for sharing this idea with us.

How Many?

Students estimate the number of items contained in a vessel using inquiry skills.

Application

Measurement • Error analysis • Volume

SI International System of Units

Theory

Students are given an opportunity to analyze a container of beans in order to develop the skills necessary to estimate the number of particles present in a volume. An error analysis may be done once the actual number of particles present in the volume is revealed and compared to the students' estimates. The activity also provides an opportunity to introduce the SI System of Measurement—in particular the unit of volume, the liter (L). The liter is the SI label representing a cubic decimeter.

Materials

Adhesive, construction (or hot glue)

Pinto beans (or similar), uncooked, 16-oz bag, 3

Rulers, metric, for each student or pair of students

Soft drink bottle, label removed, 1-L

Spring scale or balance (optional)

Tape, electrical

Preparation

Begin by filling a 1-L plastic soft drink bottle with whole pinto beans (discard all split or broken beans). Make a note of exactly how many beans are placed into the vessel. Seal the bottle cap permanently using construction adhesive or hot glue. Set the remaining beans aside—one bean per student will be required.

Safety Precautions

Although this demonstration does not pose any obvious risks, always follow laboratory safety rules.

Demonstration

Show students the bottle filled with beans and emphasize that the 1-L bottle is completely filled. Distribute one metric ruler and one whole bean to each student or team involved in this exercise. Ask students to carefully measure the size of the bean and estimate how many beans are contained in the 1-L bottle. Permit students to inspect the bottle during this process, however, provide no other assistance—let students explore on their own. Students may want to weigh the bottle and beans and compare mass. Decide if you want this to be part of the exercise.

Tabulate the data on the board and create a bar chart that summarizes the results. Reveal the actual number of beans in the bottle and ask the student who was closest to the answer to outline the procedure he or she used to make this determination. Finally, review the concepts of size and volume as well as the definition of a liter, the SI unit of volume.

Note: This start-of-the-year activity provides a great opportunity to encourage interest and enthusiasm in your physics class by rewarding students who obtain the best results with extra credit or inexpensive physics toys, such as a Rattleback or Spring Popper, which can be purchased at science supply stores. Introducing such an item also provides a spontaneous learning opportunity that students will appreciate.

Chapter 2

One-Dimensional Motion

It's All Relative

The motion of toy vehicle is viewed from two different perspectives illustrating the need for a frame of reference.

Application | Frames of reference • Motion • Fictitious forces

Theory | Motion is judged in relation to an object that is assumed to be stationary, or not moving. The object that is used for comparison is called a frame of reference. Observations of motion are made in relation to a frame of reference. A person walking to the back of a moving bus may appear to be stationary to an observer just outside the bus because the observer is using the Earth as their frame of reference, which is the most common. Passengers, however, are most likely to use the bus as their frame of reference as they walk through it. Students sitting at their desks will assume that they are not moving since they use the Earth as their frame of reference. The Earth is the most common frame of reference. However, if the Sun is used as their frame of reference, then the students are moving. Using the center of the Earth as a frame of reference, students along the 40°N parallel (roughly at the boarder between Kansas and Nebraska) will be traveling about 1300 km/hr due to the rotation of the Earth on its axis.

There are two types of frames of reference—inertial and non-inertial. Objects in an inertial frame of reference travel at a *constant* speed and in *straight* lines; that is, they do not accelerate nor rotate. In an inertial frame of reference, Newton's First Law holds true. An object in a non-inertial reference frame is undergoing linear acceleration and/or rotation. A car traveling around a curve, an object undergoing linear acceleration, and a rotating carousel all accelerate and/or rotate. Newton's First Law does not hold true for objects in non-inertial frames of reference, as objects appear to be accelerating without the presence of an appropriate force. For example, when a car suddenly accelerates, a passenger "feels" as if he or she is being pushed into the seat of the car. The feeling is quite real, but the force is said to be "fictitious." There is no force pushing the passenger backward. The passenger is *viewing* the action in a non-inertial frame of reference, whereas when viewed from the road, the passenger exists in an inertial frame of reference. In the non-inertial frame of reference the car is said to be pushing (and therefore accelerating) the passenger forward. In another example, a passenger in a car rounding a curve feels like a force is pushing her outward into the door. Again the feeling is quite real, but an outward force does not exist. When viewed from outside of the car, if the door would suddenly be removed, the passenger would not move outward along the radius but rather she would move at a tangent to the circular path. That is, because of her inertia she would have a tendency to continue on a straight path as dictated by Newton's First Law. Since the door is present, it actually pushes the passenger inward thus directing her in a circular path.

| **Materials** | AV (audio-visual) cart |

Board (chalk or marker) or wall as a backdrop

Dynamics cart and 4′ track

Toy figure (a doll small enough to ride the dynamics cart)

Note: As an alternative a low friction toy car and wooden plank may be used as a substitute for the commercial apparatus.

Safety Precautions

Although this demonstration does not pose any obvious risks, always follow laboratory safety rules while performing demonstrations.

Preparation

Place the dynamics cart on the audio-visual cart. The top shelf of the AV cart must be level. To verify if it is, place the dynamics cart on the track, if it begins to roll on its own, shim the track using pieces of paper or cardboard.

Demonstration

Place the AV cart and track in the front of a chalk or marker board. Place the dynamics cart onto one end of the track. Ask students what they think will happen to the dynamics cart if the AV cart is pushed along the floor. After discussing the various hypotheses, draw a vertical arrow on the board pointing towards the dynamics cart. (As an alternative, a piece of tape may be affixed to a wall, which will act as a backdrop. A ring stand may also be placed behind the scene as an

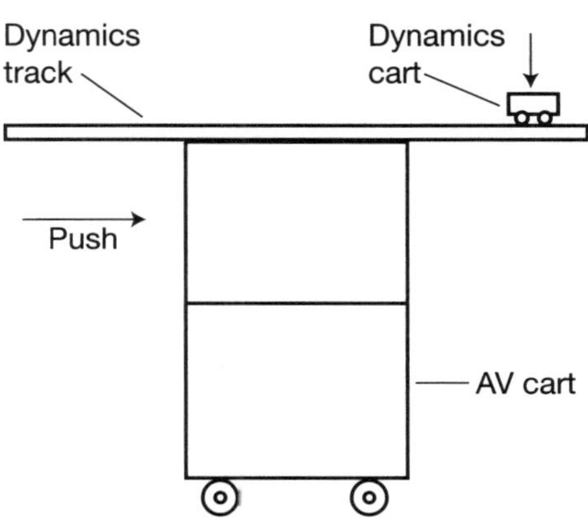

effective reference). Gently push the AV cart with a constant velocity across the room by pushing from the opposite end where the dynamics cart is situated. The dynamics cart will remain still as the roadway travels beneath it. Discuss the motion of the dynamics cart relative to the ground compared to its motion relative to the track. Emphasize the importance of identifying a frame of reference whenever observing motion.

Invite a second volunteer to assist with the demonstration. Repeat the above procedure, but this time have the second student walk alongside the moving cart. Ask this student to explain his or her relative motion to the cart. Finally, place a small toy figure on the dynamics cart and repeat the demonstration. Ask students to explain the motion of the dynamics cart with respect to the toy figure.

What's Your Slope?

Application | Graphing • Displacement • Velocity

Theory | In this demonstration, students will have an opportunity to observe an object moving at a constant rate while seeing a real-time graph of its position versus time, thus reinforcing the connection between observation and data. A position-versus-time graph of an object moving at constant velocity will show a linear or straight line relationship, $y = mx + b$. (If starting from point zero, then $y = mx$, or $d = vt$, where d is distance, v is velocity, and t is time.) The speed of the object is equivalent to the slope of the line,

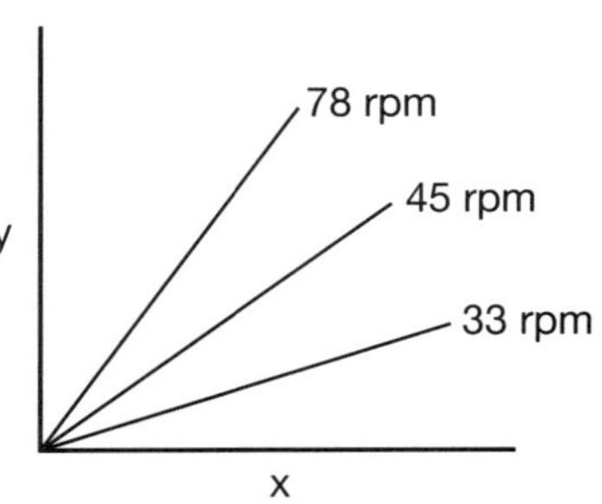

$$m = \frac{y}{x} \ \left(\text{or } v = \frac{d}{t}\right).$$

The demonstration is repeated three times with the motion sensor placed behind the moving object. Each trial is conducted with the record player set at three different speeds—producing three different slopes. A fourth trial places the motion sensor ahead of the moving object yielding a graph with a negative slope. The negative slope indicates that the object is moving closer to the sensor over time.

Materials | Dynamics cart

Tape, electrical

Hot glue

Hot glue gun

Motion sensor and computer (e.g., Pasco® or Vernier® system)

Record player with 45, 33 and 78 rpm settings*

Sewing thread spool, empty

String, 2-m

Optional: A dynamics track may be used to ensure that the motion of the cart is in a straight line. A projection device or large monitor may be used to project the image of the computer screen.

**Note:* If a record player is not available one may use a DC motor or AC motor with a rheostat.

Safety Precautions | Although this demonstration does not pose any obvious risks, always follow laboratory safety rules.

Preparation

Wrap a piece of electrical tape around the spindle of a record player in order to increase its diameter. Slide an empty thread spool onto the spindle and secure the string to a dynamics cart. Use hot glue to secure a 2-m long string to the spindle. Place a motion sensor behind the cart in line with its motion.

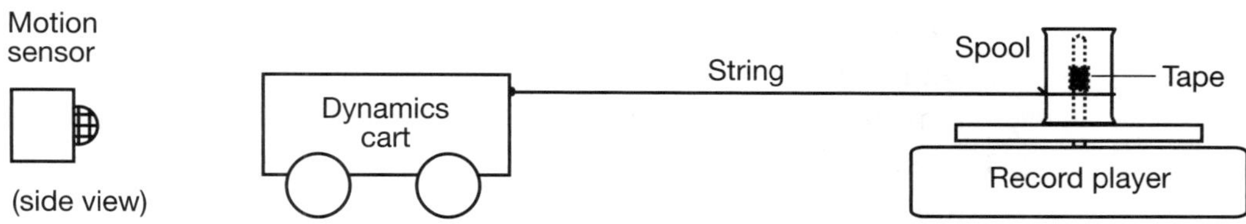

Demonstration

Set the record player at 45 rpm. Turn on the record player and simultaneously begin collecting position-versus-time data for the motion of the cart using a motion sensor connected to a computer. A straight-line graph should be produced. Ask students to explain the significance of the straight line. What does it demonstrate about the vehicle's motion?

Set the record player at 33 rpm and run the demonstration again. Ask students to compare this line to the previous one collected.

Set the record player at 78 rpm and run the demonstration once more. Ask them to interpret the graph.

Finally, position the motion sensor such that the cart is moving towards it and then ask students to interpret the results.

Change Your Slope

The motion of an object moving with a changing velocity is studied.

Application	Graphing • Velocity • Acceleration • Newton's First Law • Calculus

Theory

In this demonstration students will have an opportunity to observe an object accelerating while seeing a real-time graph of its position versus time, thus reinforcing the connection between observation and data. A position-versus-time graph of an object moving with increasing velocity will show the parabolic relationship, $y = mt^2$. The demonstration is repeated three times with the motion sensor placed behind the moving object. Each trial is conducted with different masses suspended from the string which produces three different curves. A fourth trial places the motion sensor ahead of the moving object, yielding a graph with a negative slope. The negative slope indicates that the object is moving closer to the sensor over time.

Materials

Dynamics cart

Hooked masses, e.g., 200-g, 400-g, 600-g

Motion sensor and computer (e.g., Pasco® or Vernier® system)

Pulley and pulley table clamp

String, 2-m

Optional: A dynamics track may be used to ensure that the motion of the cart is in a straight line. A projection device or large monitor may be used to project the image of the computer screen.

Safety Precautions

Although this demonstration does not pose any obvious risks, always follow laboratory safety rules.

Preparation

Secure a pulley to the edge of the demonstration table using a table pulley clamp. Tie a loop in both ends of a 2-m long string and secure one end of the string to a dynamics cart. Pass the other end of the string over the pulley and hang a 400-g mass from the string. The size of the mass necessary to move the cart is dependent upon the friction of the cart. Place a motion sensor behind the cart in line with its motion.

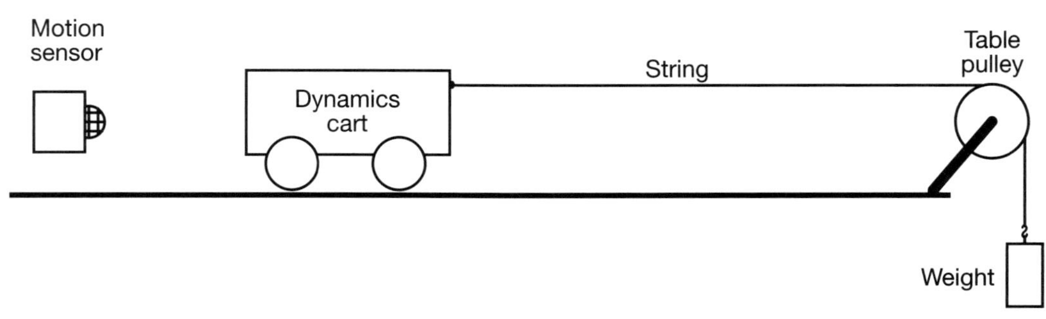

Demonstration | Direct students' attention to the apparatus and to the mechanism that influences the motion of the cart. Ask students to predict the motion of the cart. After discussing the various thoughts, allow the hanging mass to move the cart across the table. Ask students if their predictions are correct. Repeat the experiment but this time collect data using a computer-interfaced motion sensor. Explain the advantages of using technology in collecting scientific data. Begin collecting position versus-time-data of the motion of the cart using a motion sensor connected to a computer, as you allow the cart to move under the influence of the suspended mass. A curved-line graph should be produced. Ask students what is the significance of the curved line—what does it demonstrate about the vehicle's motion?

Hang a smaller mass from the string and repeat the experiment. Ask students to compare this line to the previous one collected.

Hang the largest mass from the string and repeat the experiment. Ask them to interpret the graph.

Set the software controlling the motion sensor to collect both position versus time and velocity versus time and repeat the demonstration.

Finally, position the motion sensor so the cart is moving toward it and then ask students to interpret the results.

Note: If the v versus t graph appears choppy, the rate of data collection should be decreased in the software settings.

What Are You Composed Of?

Demonstrate how force vectors are added.

Application | Vectors • Components of forces • Equilibrant force • Equilibrium

Theory | Vectors are quantities that cannot be completely described by magnitude alone. Examples are forces, velocity, acceleration, electric field strength, magnetic induction, etc. In addition to magnitude, these quantities also require a specific direction. A velocity of 25 m/s north for 10 seconds will take you to a totally different destination than a velocity of 25 m/s *east* for 10 seconds.

In this demonstration, an object, $\mathbf{W}$, is suspended in equilibrium by two ropes, $\mathbf{F}_1$ and $\mathbf{F}_2$. (In print, vectors are often represented in bold lettering.) When the ropes are arranged vertically with respect to the object, $\mathbf{F}_1 + \mathbf{F}_2 = \mathbf{W}$. $\mathbf{F}_1$ and $\mathbf{F}_2$ supply the equilibrant force that opposes $\mathbf{W}$. However, when the ropes are set at an angle, **only** their vertical components, $\mathbf{F}_{1(y)}$ and $\mathbf{F}_{2(y)}$ act as the equilibrant force. In fact, when set at an angle, $\mathbf{F}_1 + \mathbf{F}_2 > \mathbf{W}$. As $\mathbf{F}_1$ and $\mathbf{F}_2$ approach 90° as measured from the vertical, $\mathbf{F}_{1(y)}$ and $\mathbf{F}_{2(y)}$ approach infinity. Since the object does not move horizontally for all angles, $\mathbf{F}_{1(x)} = -\mathbf{F}_{2(x)}$. When each rope is analyzed separately, the components of the vector $\mathbf{F}_1$, $\mathbf{F}_{1(x)}$ and $\mathbf{F}_{1(y)}$ may be added together using the Pythagorean theorem: $\mathbf{F}_1{}^2 = \mathbf{F}_{1(x)}{}^2 + \mathbf{F}_{1(y)}{}^2$.

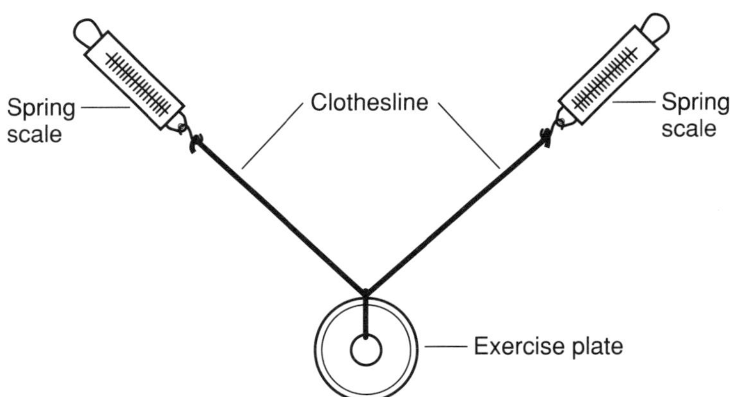

Materials | Board, magnetic, or chalkboard

Clothesline cord, 1.5 m

Exercise plate, round, 5 lb

Meter stick

Protractor, large

Ring stands, large, 2

Spring scales, 20 N, 2 (preferably magnetic marker/chalkboard demo scales)

Utility clamps, 2

<table>
<tr><td>Safety
Precautions</td><td>Although this demonstration does not pose any obvious risks, always follow laboratory safety rules.</td></tr>
<tr><td>Demonstration</td><td>Suspend a 5-pound (22 N) exercise plate with the cords arranged vertically. Each spring scale should register approximately 11 N.</td></tr>
</table>

If available, suspend the apparatus from magnetic board demonstration scales; otherwise, suspend the spring scales from two ring stands. Mark the three lines of force on the marker board. Record the magnitude of each force next to the line including the downward force of the exercise weight. Draw the appropriate components of the vectors with respect to each line of force. Show how the two *y*-components calculated from the readings on the scales will always add up to 22 N—the equilibrant force of the system.

Slowly pull the vertically aligned cords away from each other to demonstrate how the tension in the cord, as read on the scales, increases as the angle with respect to the vertical increases. Ask students why the force is greater when the cords are at an angle.

Pull the cords away from each other so each is approximately 50° measured from the vertical. Measure the tension in each cord. It should be about 17 N. Have students calculate the vertical component of each cord and then ask them to explain the result.

What Are You Composed Of? Revisited

Demonstrate how force vectors are added.

Application	Vectors • Components of forces • Newton's Second Law • Friction
Theory	In this demonstration a block, **M**, placed on a level surface is pulled by a horizontally aligned rope with a continuous force, **F**, causing it to move at a constant velocity. Since the force is in the same direction as the motion and the block does not accelerate, this force represents the minimum force needed to move the block. However, when the rope is set at an angle with respect to the horizontal surface, a greater force is needed to move the block. The greater the angle—the greater the force. Only the horizontal component of **F**, $\mathbf{F}_x$ actually causes horizontal motion. The vertical component, $\mathbf{F}_y$ opposes the downward pull of gravity, thereby reducing the amount of friction between the block and the horizontal surface. The components of the vector $\mathbf{F}_x$ and $\mathbf{F}_y$ may be added together and yield the vector **F** using the Pythagorean theorem: $\mathbf{F}^2 = \mathbf{F}_x{}^2 + \mathbf{F}_y{}^2$.
Materials	Spring scale, 20 N, demonstration String, 30–40 cm Wood block with a screw eye on one end, $4'' \times 4'' \times 8''$ *Optional:* Use several bricks or other heavy objects to load onto the block.
Safety Precautions	Although this demonstration does not pose any obvious risks, always follow laboratory safety rules while performing demonstrations.
Preparation	Tie a loop in both ends of a 30- to 40-cm string. Tie one end of the string to the block, and the other to the demonstration spring scale.
Demonstration	Use the scale to pull the cart across the demonstration desk. Start with the scale and string parallel to the benchtop and record the minimum force needed to move the block at a constant velocity. Ask students what would happen if the string were pulled at an angle to the table. After discussing the various ideas, repeat the process with the scale raised to different angles. You might want to resolve these force vectors into their *x* and *y* components.

You Go This-a-Way and I'll Go That-a-Way

The relative motion of one moving vehicle is compared to that of another.

Application | Relative velocity • Vectors • Frames of reference

Theory | Motion is judged in relation to an object that is assumed to be stationary, or not moving. The object that is used for comparison is called a frame of reference. Observations of motion are made in relation to a frame of reference. A person walking to the back of a moving bus may appear to be stationary to an observer just outside the bus because the observer is using the Earth as his frame of reference, which is the most common. Passengers, however, are most likely to use the bus as their frame of reference as they walk through it. Students sitting at their desk will assume that they are not moving since they use the Earth as his frame of reference. The Earth is the most common frame of reference. However, if the Sun is used as their frame of reference, then the students are moving. Using the center of the Earth as a frame of reference, students along the 40th parallel will be traveling about 1300 km/hr due to the rotation of the Earth on its axis.

Materials | Demonstration table, $\approx$ 3 m long

Furring strip, 30 cm (or width of paper)

Hot glue

Hot glue gun

Paper, wax, wrapping or butcher, 3 m

Slow-moving, constant-velocity toy vehicles, 2

Note: If a 3 m long demonstration table is not available, the demonstration may be performed on the floor.

Safety Precautions | Although this demonstration does not pose any obvious risks, always follow laboratory safety rules.

Preparation | Glue a furring strip to the edge of one end of a sheet of wax paper, heavy wrapping paper or butcher paper. Alternatively, a 30-cm ruler and tape may be used.

The toy vehicle must be slow moving for this demonstration. A speed of about 3–5 cm/s is ideal. To lower the speed you will need to decrease the voltage powering the toy's motor. If necessary, replace one of the batteries with a dowel of equal size or a dead battery wrapped with aluminum foil.

Demonstration | Discuss and then demonstrate the different ways of describing the relative motion of each vehicle in the following situations:

Moving in Same Direction: $V_1 = V_2$

Position two vehicles side by side, moving in the same direction.

Moving in Opposite Directions: $V_1 = -V_2$

Position two vehicles side-by-side, moving in opposite directions.

One Moving, One Not: $V_1 = 0$; $V_2 > 0$

Position two vehicles side by side; one is stationary while the other is moving.

Road Is Moving: $V_1 > 0$; $V_2 = 0$ (relative to its surface)

Lay the heavy wrapping paper on the surface of the demonstration table so about 1.5 m is on the table, with the rest of the paper hanging over the edge. Position one vehicle on the edge of the paper about 1 m away and facing the furring strip. Place a second vehicle on the table surface, in line with the first vehicle facing in the same direction. As the second vehicle moves, powered by its batteries, gently pull the paper sheet with the non-moving vehicle on it.

Car and Road Are Moving: $V_2 = 2V_1$

Position one vehicle on the edge of the paper about 1 m away and facing the furring strip. Place a second vehicle on the table surface, in line with the first vehicle facing in the same direction. Set both vehicles in motion while pulling on the sheet with a velocity equal to the vehicle traveling on the table surface.

Car and Road Are Moving in Opposite Direction: $V_2 = -2V_1$

Position one vehicle on the edge of the paper about 1 m away and facing the furring strip. Place a second vehicle on the table surface, in line with the first vehicle facing in the opposite direction. Set both vehicles in motion while pulling on the sheet with a velocity equal to the vehicle traveling on the table surface.

Falling Together

The motion of objects falling in air is compared to the same objects falling in a vacuum.

Application Gravity • Free fall • Acceleration due to gravity • Air resistance

Theory During the 1600s, Galileo studied the rate at which objects fall by analyzing the rate at which balls roll down an inclined plane. He was able to determine that free-falling objects fall a vertical distance which is proportional to the square of the time of fall. He concluded that the acceleration due to gravity might be considered constant for a body falling a short distance. Assuming that there is no air resistance, objects on Earth fall at a rate of 9.8 m/s^2. During the Apollo 15 mission, astronaut David Scott dropped a feather and hammer on the moon illustrating what Galileo predicted would happen. A video of the experiment may be found on the Internet.

When two objects fall through air, the object with the greater air resistance will reach terminal velocity and will appear to fall at a slower rate than the object with less air resistance. If the air is removed from the system, the two objects will fall at the same rate since the resisting medium has been removed.

Materials Penny–Feather Tube* (originally called a Guinea–Feather Tube)

Vacuum pump (a Nalgene® hand-operated vacuum pump is sufficient)

Note: If a Penny–Feather Tube is not available, one may be fashioned from the following materials:

Glass tube, small 10 cm long

Paper confetti, red, 1 large piece the size of a penny

Penny

Pinch clamp

Rubber stopper, to fit plastic tube

Rubber stopper, one-hole, to fit plastic tube

Tube, plastic, 1 m long and 2–3 cm I.D. with a thickness of 0.5 cm

Tygon® tubing

Safety Precautions Wear safety goggles while performing this demonstration. Remove only enough air from the tube necessary to perform the demonstration to avoid implosion. The tube must be free of scratches to prevent implosions. Reintroduce air back into the tube prior to storage. Always follow laboratory safety rules while performing demonstrations.

<table>
<tr><td>Preparation</td><td>To prepare a homemade Penny–Feather Tube, insert a rubber stopper into one end of a large plastic tube. Place one piece of confetti and a penny into the tube. Insert a small piece of glass tubing into a one-hole rubber stopper and insert this assembly into the open end of the large tube. Caution: Use glycerin if needed to lubricate the insertion of the tube. Slide a piece of Tygon tubing over the small glass tubing.</td></tr>
</table>

Demonstration

Hold the penny tube in a vertical position. Shut the stopcock of the commercial tube (or pinch-clamp the Tygon tubing in the homemade tube). Invert the tube. The feather or confetti will fall at a slower rate than the penny.

Evacuate the tube of air. Shut the stopcock of the tube. Invert the tube. The two objects will fall at the same rate.

Note: The motion of the objects in this demonstration is rather fast, allowing only a limited time for observation. Consider videotaping the demonstration and reviewing the tape in slow motion (frame by frame).

Disposal

None required.

Reference

Hammer and Feather; Apollo 15 Lunar Video Library: NASA; Retrieved November, 2008. http://www.hq.nasa.gov/alsj/a15/video15.html#closeout3

Acceleration due to gravity is shown to be constant for a body falling a short distance.

Application	Gravity • Free fall • Acceleration due to gravity • Galileo
Theory	During the 1600s, Galileo studied the rate at which objects fall by analyzing the rate at which balls roll down an inclined plane. He was able to determine that free-falling objects fall a vertical distance which is proportional to the square of the time of fall. He concluded that the acceleration due to gravity might be considered constant for a body falling a short distance. Assuming that there is no air resistance, objects on Earth fall at a rate of $9.8 \ m/s^2$.

In this demonstration, nuts are strategically aligned at distances in ratio to 1, 4, 9, 16, 25, etc., or the squares of the natural numbers, 12, 22, 32, 42, 52, etc. When released all at once, the nuts will collide with the floor with equal time intervals, even though distance intervals are unequal. A second string of nuts with the nuts arranged at regular intervals is then dropped in the same fashion. In this case, the nuts will collide with the floor with decreasing time intervals.

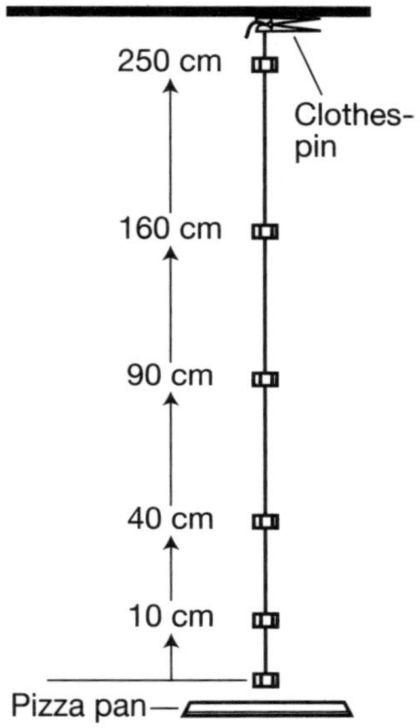

Materials

Clothespin

Nuts, ½″, 12

Pizza pan

Soup cans, empty, or paper towel tubes, 2

String, nylon, thin

Safety Precautions

Although this demonstration does not pose any obvious risks, always follow laboratory safety rules.

Preparation

Tie six nuts to a piece of thin nylon cord at intervals measured from the end as 10 cm, 40 cm, 90 cm, 160 cm, 250 cm, etc., as indicated in the diagram.

Prepare second string of nuts, but with the nuts spaced every 50 cm (i.e., at regular intervals).

Demonstration

Secure a clothespin from the ceiling of the demonstration area. Suspend the apparatus from the clothespin over an inverted pizza pan.

Ask students to close their eyes to eliminate visual distractions.

Open the clothespin and allow the string of nuts to fall onto the pizza pan. Instruct students to listen to the impact that the nuts make with the pizza pan. Students will notice that the sounds are evenly spaced.

Repeat the demonstration using the string with the nuts arranged at regular intervals and discuss the difference in time.

Note: To prevent tangling, store the string of nuts by wrapping them around separate cans or paper towel tubes.

Reference

Blasi, Rocco C. Ed. *Physics Fun and Demonstrations with Professor Julius Sumner Miller;* Central Scientific: Chicago, IL, 1974; p 152.

Air resistance affects the rate of fall of some objects.

Application | Gravity • Acceleration • Air resistance • Terminal velocity

Theory | Air resistance, a form of friction, is a force exerted by air on a moving object. It acts in the opposite direction of an object's motion and depends on the mass, shape and speed of the object. Of these three variables, the shape of an object has the greatest influence on the magnitude of air resistance. Air resistance may have little effect on a dense spherically-shaped object, however, if this same object were reshaped into a flat disk, the difference in air resistance would be significant. A goal in automobile design is to reduce air resistance in order to increase speed, while a parachute is designed to maximize the amount of air resistance.

As an object falls, the force of gravity causes the object to accelerate. If no other force acted on the object, the graph of the position versus time would be parabolic, $y = \frac{1}{2} gt^2$. Air resistance, however, provides a second yet opposing force acting on the object. Air resistance increases as the object accelerates. When the upward force of air resistance equals the downward force of gravity, the object stops accelerating and reaches terminal velocity. Terminal velocity is the maximum velocity attained by a falling object. The position–time graph of a falling object under the influence of air resistance will initially be curved and then approach a linear relationship as the terminal velocity is reached.

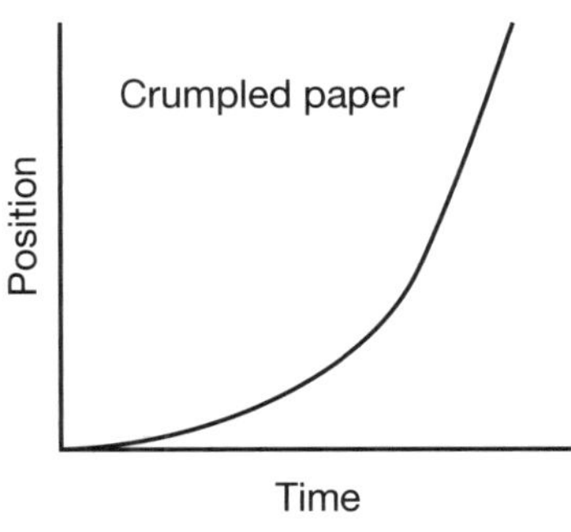

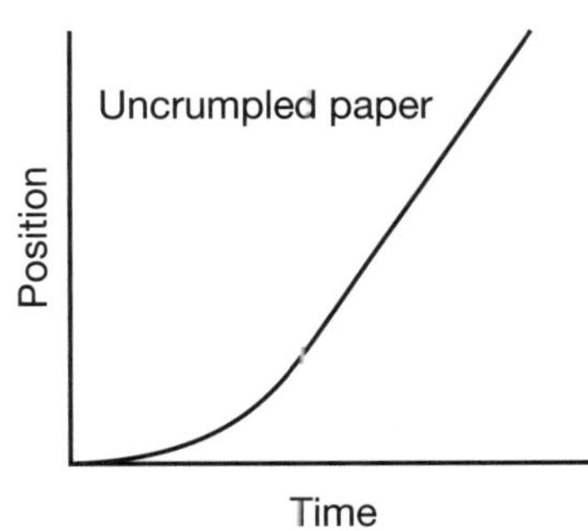

Materials | Coffee filters, large, round, 3 or more

Motion sensor and computer (e.g., Pasco® or Vernier® system)

Optional: A projection device or large monitor may be used to project the image of the computer screen.

Safety Precautions | Although this demonstration does not pose any obvious risks, always follow laboratory safety rules.

Demonstration | Crumple up one coffee filter into a tight ball. While standing on top of the demonstration table, drop the crumpled ball and a second filter at the same time from the same height towards the floor. The crumpled ball should fall straight down while the sheet will float downward. Point out that the two filters have an identical mass. Discuss how the shape of an object affects its movement through air or water.

Ask students to sketch in their notebooks what the position versus time graphs of the motion would look like. Repeat the experiment, one filter at a time, dropping the filters over a motion sensor directed upward. Compare the graphs.

Finally, demonstrate the effect of mass on air resistance by dropping two or more stacked coffee filters versus one filter.

The acceleration of an object due to gravity is independent of its mass.

Application

Gravity • Free fall • Acceleration due to gravity • Terminal velocity

Theory

Many students incorrectly assume that the heavier an object the faster it falls. Although the downward acceleration of an object may be affected by its air resistance, its rate of fall is independent of its mass. Two objects of similar shapes with dissimilar mass will fall reasonably together in the presence of air. This demonstration uses spheres as they encounter the least air resistance of common shapes. The spheres must have a large enough mass in order to make the affects of air resistance negligible. If dropped from a great height, such as a tall building, the affect of air resistance will become apparent since more time is available to approach or reach terminal velocity.

Materials

Aluminum sphere, 1″ (e.g., pendulum bob)

Lead sphere, 1″

Ping Pong ball

Pizza pan

Safety Precautions

This demonstration is relatively hazard free, however, lead objects should be handled with care due to their toxicity. Wash hands after handling lead objects.

Demonstration

Show students the two spheres and ask them to describe them. Most likely students will notice the pronounced difference in mass between the two objects ($D_{Pb} = 11.3$ g/cm^3; $D_{Al} = 2.7$ g/cm^3). Then ask them to predict which would strike the ground first if released from the same height at the same time. After discussing the various ideas, hold the balls high and drop them at the same time onto a pizza pan placed on the floor. Direct students' attention to the fact that the balls make one sound as both hit the ground together.

In contrast, drop the aluminum ball together with a Ping-Pong ball. Discuss why the Ping-Pong ball falls at a slower rate than the denser (more compact) aluminum ball.

Note: The motion of the objects in this demonstration is rather fast resulting in a limited time for observation. Consider videotaping the demonstration and reviewing the tape in slow motion (frame by frame).

Money from the Sky

Students measure their reaction times by attempting to catch a falling dollar bill.

Application

Gravity • Acceleration due to gravity • Reaction time

Theory

Human reaction times can be as quick as 100 milliseconds (0.1 sec). Reaction time depends upon a number of physiological factors, including age, state of health, and even the distance the nerve impulses must travel between the muscles and brain. In this demonstration, a student attempts to catch a falling dollar bill. It takes 0.13 sec for the bill to fall half its length (7.7 cm). The process begins with the eye sending a signal to the brain of what it sees. The average time for an impulse to travel from eye to brain to fingers is at least 0.14 seconds. The demonstration is then repeated using a meter stick. The distance that the meter stick falls will be recorded and used to calculate the student's reaction time using the equation:

$$d = \frac{1}{2} gt^2 \quad \text{or} \quad t = \sqrt{\frac{2d}{g}}$$

where d is the distance the meter stick falls and $g = 9.8 \text{ m/s}^2$.

Materials

Dollar bill, crisp (new)

Meter stick

Safety Precautions

Although this demonstration does not pose any obvious risks, always follow laboratory safety rules.

Demonstration

Catching a Dollar Bill

Offer students a chance to earn a dollar by inviting one student to serve as an assistant. Tell the assistant that if she can catch a falling dollar bill then she can keep it. Hold a crisp dollar bill vertically from one end and ask the student to position her open fingers on either side of the middle of the bill. Instruct the student that when the bill is released, she is to grab the bill using only her fingers. She may not move her hand down with the bill. Students will find the objective impossible to achieve since the bill is not long enough with respect to the students' reaction time.

Measuring Reaction Time

Repeat the demonstration using a meter stick and determine how far the meter stick falls from the point of release to the point of being caught. Calculate the reaction time of the student using the equation for free fall: $d = \frac{1}{2} gt^2$.

References

Meiners, H. F. *Physics Demonstration Experiments;* The Ronald Press Company: New York, 1970, Vol I.

Kraus, David. *Concepts in Modern Biology;* Globe Book Company: New York, 1969, p 170.

A steady stream of falling water droplets are "frozen" using a stroboscope to illustrate the acceleration of objects in free fall.

Application | Gravity • Free fall • Acceleration due to gravity

Theory | A stroboscope is a light which is similar to a photographic flash unit but that can produce flashes at a rate as high as 10,000 or more per second. When used in a darkened room, the flash of a stroboscope can be used to "freeze" the movement of an object. Stroboscopes are commonly used at social dances to create an interrupted motion effect as people move across the dance floor. The same technique may be employed to study the rate of motion of an object. The stroboscope reveals the position of a moving object as it moves through the dark. The speed of an object moving at a constant velocity may be determined by measuring the change in position between revealing flashes and dividing it by the rate (i.e., time) of the flash unit. The changing position of the object may be preserved by taking a photograph with the film being exposed during the entire movement resulting in a multiple exposure effect. Harold "Doc" Edgerton (1903–1990) of the Massachusetts Institute of Technology developed techniques, called ultra-high speed and stop-action photography, that utilized the stroboscopic effect to analyze moving objects. The *Balloon and the Bullet* and the *Milk Drop Coronet* are two of his most widely seen photographs. In this demonstration, free-falling water droplets are "frozen" in space using a stroboscope in order to view the increasing distance between successive drops as the drops move away from the tip of the buret. The increase in distance demonstrates that the drops are accelerating.

Materials | Bowl, cup or beaker

Buret

C-clamp

Clamp

Meter stick

Ring stand, tall

Stroboscope

Safety Precautions | Speak to your school nurse to verify if any of your students have a history of seizures since the flashing of a stroboscope has been known to initiate a seizure in some people.

Preparation | Attach a buret to a stand and fill it with water. Use a C-clamp to secure the base of the ring stand to a table such that the buret is hanging over the edge of the table. Place a bowl, cup or beaker on the floor directly under the tip of the buret. A large distance between the tip of the buret and the bowl will provide the most dramatic observations.

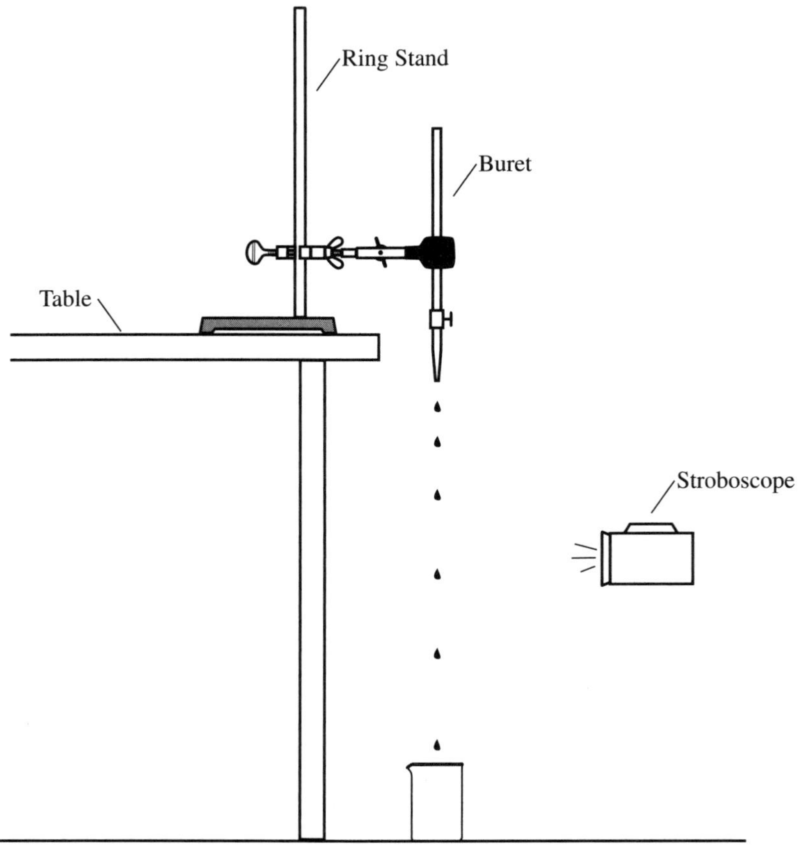

Demonstration

Turn the stopcock of the buret and adjust the flow of water such that it leaves the buret at a rate of approximately 3–5 drops per second. Turn on the stroboscope and then darken the room. Adjust the stroboscope such that its flashes match the rate of the drops leaving the burette causing the position of the drops to appear "frozen" in air. Direct students' attention to the fact that the distance between successive drops increases as the drops move away from the tip of the buret. The increase in distance demonstrates that the drops are accelerating.

A meter stick may be used to measure the distance between the drops in order to verify the acceleration due to gravity.

Reference

Based on an exhibit at the Boston Museum of Science, May 2005.

Chapter 3

Two-Dimensional Motion

Why Did the Boat Cross the River?

The motion of a vehicle under the influence of two forces is observed and analyzed.

Application | Two-dimensional motion • Vectors • Frame of reference

Theory | A boat crossing a river is under the influence of two forces—the force resulting from the action of the motor as well as the force of the current. The actual velocity of the boat is a vector combination of the velocities resulting from the two forces. If the boat is directed at a right angle to the current, the velocity may be determined using the Pythagorean Theorem: $\mathbf{v}^2 = \mathbf{v}_x^2 + \mathbf{v}_y^2$. If the boat is directed at other angles, the velocities must first be broken down into their respective x and y components. The length of the paper used in this demonstration provides a visual effect for students to imagine that they are looking at a boat crossing a moving stream. Boats and airplanes encounter outside forces due to wind and water currents that influence the actual direction requiring the captain to compensate by steering toward these forces.

Materials | Bamboo skewers, 3

Demonstration surface, $\approx$ 3 m long

Furring strip, 30 cm (or width of paper)

Hot glue

Hot glue gun

Index cards, 3

Permanent marker

Rubber stoppers, 1-hole, #6 or larger, 3

Toy vehicle, battery-operated (slow-moving and constant velocity)

Wax, wrapping or butcher paper, 3 m

Note: If a 3-m long demonstration table is not available, the demonstration may be performed on the floor.

Safety Precaution | Although this demonstration does not pose any obvious risks, always follow laboratory safety rules.

Preparation | Glue a furring strip to the edge of one end of a sheet of heavy wrapping or wax paper. Alternatively, a 30-cm ruler and tape may be used.

The toy vehicle must be slow-moving for this demonstration. A speed of about 3–5 cm/s is ideal. To lower the speed, you will need to decrease the voltage powering the toy's motor. If necessary, replace one of the batteries with a dowel of equal size or a dead battery wrapped with aluminum foil.

Prepare three reference flags by securing a bamboo skewer in the hole of a #6 (or larger) rubber stopper. Glue a flag, made from an index card onto which a number has been drawn using a permanent marker. These flags can be used for other demonstrations as reference points also.

Demonstration

Lay the heavy wrapping paper on the surface of the demonstration table so about 1.5 m is on the table, with the rest hanging over the edge. Position a constant velocity toy vehicle on the edge of the paper about 1 m from the furring strip. The car should be facing parallel to the furring strip. Ask students to describe the movement of the vehicle as it crosses the paper road. After a short discussion, confirm their predictions by allowing the vehicle to move across the paper.

Now ask students to pretend that the vehicle is a boat and ask them to describe its motion if it were crossing a river having a current moving at a right angle to the boat. Use a reference flag to mark the initial position of the vehicle with respect to the table. The other two flags can be used to mark students' predictions. After a short discussion, test students' hypotheses by allowing the vehicle to move across the paper as you gently pull on the furring strip to simulate the moving river.

In addition, the motion and velocity of the boat may be explored by sending the vehicle in the direction of the moving paper, showing how the current can increase the velocity of the vehicle. Finally, the vehicle may be sent against the direction of the moving paper, showing how the velocity may be reduced. The velocity may be reduced to zero by matching the speed of the moving paper with the speed of the vehicle. The vehicle may be given a negative velocity by pulling the paper faster than the speed of the vehicle. Consider discussing the different frames of reference (i.e., car versus river or river versus car) with each of these examples.

Drop the Dice

A quick and easy way to show that horizontal velocity has no effect on rate of fall.

Application

Projectile motion • Two-dimensional motion • Free fall

Theory

The concept that the vertical and horizontal components of the velocity of a projectile are *independent* of each other is often a difficult concept for students to grasp. An excellent way to demonstrate this concept is by utilizing a commercial device often called a Second Law of Motion Apparatus. This device uses a spring-loaded launcher to drop two balls: one vertically (in the absence of a horizontal velocity) while another ball is projected with an initial horizontal velocity and free to drop. In this demonstration a pair of dice is launched from a table using an ordinary ruler in such a way that one die has a significantly larger horizontal velocity than the other. Despite this, the pair of dice strikes the floor at the same time, indicating that the time to reach the floor is independent of the horizontal velocity. This demonstration is often performed using coins or balls. Dice are more visible than coins, however, and do not roll away like balls would upon landing.

Materials

Pair of dice

Ruler, flexible, 12-inch

Optional: Second Law of Motion Apparatus and a video camera and projection device

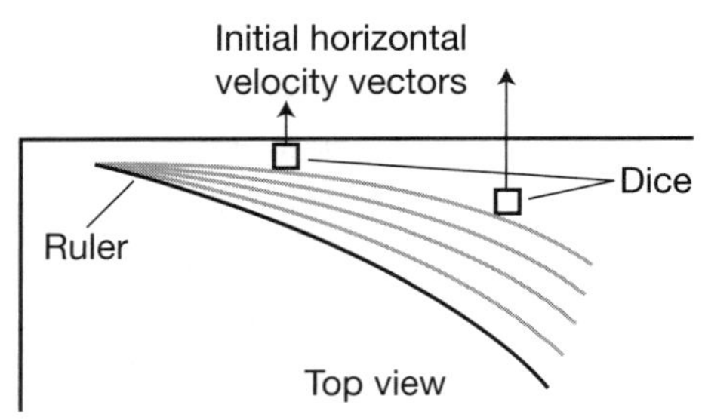

Safety Precautions

Although this demonstration does not pose any obvious risks, always follow laboratory safety rules.

Demonstration

Place the pair of dice near the edge of the demonstration table, with one die closer to the edge than the other. Use the ruler to flick the dice off of the table simultaneously. Tell students to listen carefully for the sound of the dice hitting the floor. They should hear a single clack as the dice hit the floor simultaneously. Emphasize the fact that the dice must have left the table with different horizontal velocities since the time of fall is the same, yet the horizontal distances from the table are different.

To effectively perform this demonstration, be sure to practice the procedure a few times since it requires some minimum skill.

Note: The motion of the objects in this demonstration is rather fast resulting in a limited time for observation. Consider videotaping the demonstration and reviewing the tape in slow motion or frame by frame.

A block of wood is launched horizontally to illustrate how it immediately begins to free-fall upon its release.

Application	Projectile motion • Two-dimensional motion • Free fall
Theory	A projectile launched horizontally immediately begins and then continues to fall freely due to the influence of gravity. This is often a difficult concept for students to understand. Students will incorrectly assume that if the projectile moves fast enough, then the horizontal component of motion will somehow affect vertical movement. This misconception is regularly reinforced in motion pictures and in cartoons. Horizontal velocity only affects horizontal displacement!

Theory (continued)

In this demonstration a block of wood is struck by a second block, which causes the first block to move off of a table. The intention is to make the first block land on a neighboring table of equal height. No matter how fast the block is launched, it cannot land on the neighboring table because it immediately begins to fall once it leaves the first table.

To ensure the first block is launched horizontally, it is important that all surfaces are smooth. The second block of wood is larger to increase the momentum of the system. The higher center of gravity of the second block will also assist in preventing the first block from becoming airborne before it leaves the table.

Materials

Movable tables with the exact same height, 2

Wooden block, $2'' \times 4'' \times 6''$, with sharply cut ends and with surfaces sanded smooth

Wooden block, $4'' \times 4'' \times 6''$, with sharply cut ends and with surfaces sanded smooth

Note: The height of some tables may be adjusted at the base of the table leg.

Safety Precautions

Although this demonstration does not pose any obvious risks, always follow laboratory safety rules.

Demonstration

Place two tables, having the exact same height, about 30 cm apart. Place a $2'' \times 4'' \times 6''$ wooden block flat on the edge of one table with the smaller face ($2'' \times 4''$) pointed towards the other table.

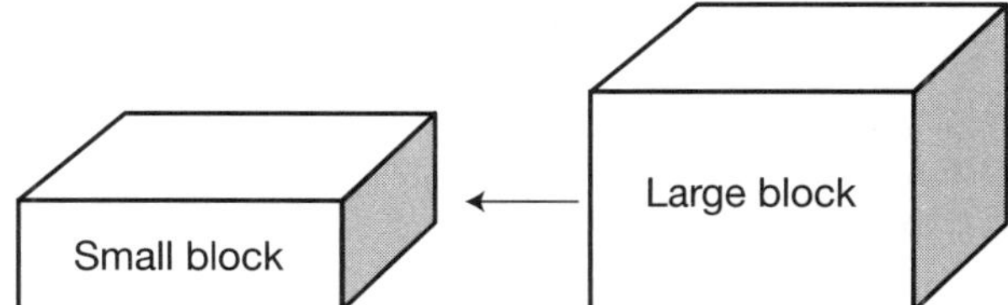

Ask students to predict whether it is possible to strike the smaller block sharply enough with the larger block in a horizontal direction, so that the smaller block will be able to land on the second table. Upon discussion, quickly slide the larger of the two blocks across the table and sharply strike the smaller block to cause it to slide toward the second table. Try different speeds. No matter how fast the smaller block moves, it will always strike the leading edge of the second table and then fall between the two tables or flip in the process.

The demonstration can be concluded with a short review or screening of movie scenes that reinforce the misconception of the effect of horizontal velocity on vertical motion. Students may be familiar with frequently recurring scenes of the cartoon character Wile E. Coyote running off a cliff and not falling until he realizes that he is no longer on the cliff. Movies often depict cars leaving cliffs with an upward jump or an apparent upward force.

Note: The motion of the objects in this demonstration is rather fast, resulting in a limited time for observation. Consider videotaping the demonstration and reviewing the tape in slow motion or frame by frame.

Parabolic Garden Hose

A water hose is used to illustrate the parabolic path of a projectile.

Application | Projectile motion • Two-dimensional motion • Parabolic trajectory

Theory | An example of the parabolic path followed by the two-dimensional motion of an object may be illustrated by observing the flight of a baseball or arrow curving through air. Of course, air resistance must be neglected, but for all practical purposes the path appears near parabolic. In this demonstration, the parabolic trajectory of two-dimensional motion is illustrated using a thin stream of water. Unlike the baseball or arrow, the launch angle of the stream may be quickly altered, showing that the maximum range of the projectile is achieved if launched at a 45° angle. The continuous stream of water makes the parabolic trajectory visible. It can also be shown that every point in the range may be achieved from two different launch angles—for example, 30° and 60°.

Materials | Bamboo skewers, 5

Eye-dropper tube, glass , >5 cm long

Hose, latex or rubber, 1-m

Hot glue

Index cards, 5

Permanent marker

Protractor, large

Protractor, large (chalkboard-type)

Rubber stopper, one-hole, (#3, 4 or 5)

Rubber stoppers, one-hole, (#6 or larger), 5

Ring stand

Utility clamp, swivel-type

Wallpaper trough (available at home hardware stores)

Water faucet with hose connector

Note: If a garden hose and outdoor faucet are accessible, the demonstration may be easier to perform due to the minimal setup and cleanup involved. A sextant, such as described in *How High?* found on page 2, may be permanently attached to the side of a hose nozzle dedicated to this demonstration.

Safety Precautions | Although this demonstration does not pose any obvious risks, always follow laboratory safety rules.

Preparation | Insert a long glass eye-dropper tube into a one-hole rubber stopper, with the dropper end protruding. Center the rubber stopper on the tubing. Attach a 1 meter long piece of latex or rubber tubing to one end of the glass tubing–stopper assembly.

Secure the glass tubing–stopper assembly to a swivel-type utility clamp attached to a ring stand. Secure the free end of the rubber tubing to the hose connector of a water faucet. Set a wallpaper trough lengthwise in the path of the water that will be emitted from the glass tubing–stopper assembly.

Prepare three reference flags by securing a bamboo skewer in the hole of a #6 (or larger) rubber stopper. Glue a flag, made from an index card onto which a number has been prominently drawn on using a permanent marker. These flags can be used for other demonstrations as reference points also.

Demonstration | Begin with a discussion regarding the path a projectile takes when launched at an angle. Ask students what, if any, is the "ideal angle" that would provide the greatest horizontal distance traveled by the projectile.

Turn on the water faucet slowly until a steady stream of water is emitted from the assembly. The water should land in the trough at all angles. Use a large protractor to measure the angles. Beginning at about 10°, adjust the angle of the assembly and note how the horizontal distance at first increases as the angle of 45° is approached, and then decreases again at larger angles. Emphasize that there appear to be angles above and below 45° that provide the same horizontal range. Locate the range of the water stream for angles 30°, 40°, 45°, 50° and 60°. Place reference flags along the side of the trough to mark the spots where the water stream lands for each angle.

Darts are launched at various angles using a spring-loaded dart gun.

Application | Projectile motion • Two-dimensional motion • Parabolas

Theory | An example of the parabolic path followed by the two-dimensional motion of an object may be illustrated by observing the flight of a baseball or arrow curving through air. Of course, air resistance is always present, but for all practical purposes, it can be ignored so the path appears near parabolic. In this demonstration, the parabolic path of two-dimensional motion is illustrated using a spring-loaded dart gun. Unlike the baseball or arrow, the angle of the dart may be easily altered with successive launches showing that the maximum range of the projectile is achieved if launched at a 45° angle. It can also be shown that every point in the range may be achieved from two different launch angles—for example, 30° and 60°.

Materials | Bamboo skewers, 5

Dart gun

Fishing line or thread

Hot glue

Hot glue gun

Index cards, 5

Permanent marker

Protractor, plastic

Rubber stoppers, 1-hole, (#6 or larger), 5

Soda straw

Tape measure or meter stick, 5-m

Washer, ½″, or small fishing weight or sinker

Safety Precautions | Use safety goggles when working with traditional rigid-type dart guns, as they have been known to cause severe eye injury. The rigid darts have been known to spontaneously launch from the loaded gun without warning. Always point the gun away from your audience and yourself.

Preparation | Construct a sextant by hot gluing a plastic soda straw to the base of a plastic protractor. Tie a washer or fishing weight or sinker to the end of apiece of fishing line or thread. Use hot glue to secure this line to the middle of the straw. Hot glue this

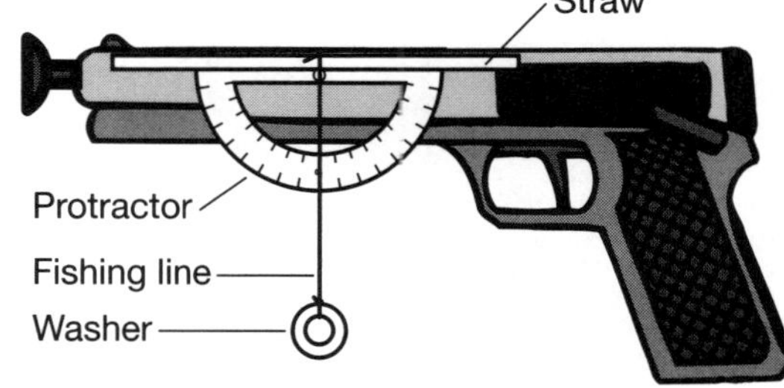

apparatus to the barrel of the dart gun. Prepare three reference flags by securing a bamboo skewer in the hole of a #6 (or larger) rubber stopper. Glue a flag, made from an index card onto which a number has been prominently drawn on using a permanent marker. These flags can be used for other demonstrations as reference points also.

Demonstration

Clear a 5-m path on the classroom floor or move into an area such as a hallway to perform this demonstration. While holding the dart gun at floor level, launch the dart directly upward a few times to demonstrate that the dart gun launches the dart relatively the same distance each time. If the classroom ceiling is not high enough for this part of the demonstration, the dart can be launched horizontally a number of times to determine the average range. Be sure that the dart is launched from the same spot using the same angle each time.

Once it has been established that the dart gun behaves consistently, ask students how the gun should be aimed, with the handle touching the floor, in order for the dart to travel as far as possible. Discuss and test all hypotheses. Most likely some students will realize that the ideal angle for maximum range is 45°.

Ask students what happens if the launch angle increases beyond 45°. After discussion, search out the relationship between the launch angles below and above 45°. You should find, within a reasonable approximation, that there are two launch angles for each point in the range. Locate the range of the dart for angles 30°, 40°, 45°, 50° and 60°. Place reference flags along the floor to mark the spots where the darts land.

A ball is thrown up and down while the demonstrator moves on roller skates illustrating the parabolic path of a projectile.

Application

Projectile motion • Two-dimensional motion

Parabolic trajectory • Frame of reference

Theory

The horizontal component of two-dimensional motion is independent of the vertical component. In this demonstration, a small round plush toy is thrown up and down while the demonstrator moves about the classroom on roller skates at a constant speed. The toy, which is in the hands of the demonstrator, is traveling with the same horizontal speed as the demonstrator. The demonstrator then provides an upward force on the toy, which gives the toy an initial vertical velocity causing it to move upward. The toy decelerates as it moves upward due to the downward effect of gravity and then accelerates as it falls back down towards the demonstrator's hand. The horizontal motion of the toy is in sync with the horizontal motion of the demonstrator's hand the entire time and may be explained by Newton's First Law—an object in motion (in this case horizontal) will stay in motion unless an unbalanced force acts upon it. In this case, air resistance is negligible and therefore may be ignored. The resulting path of the toy is parabolic.

Materials

Plush toy, small, round, or similar object

Roller skates or roller blades

Football helmet or other sport helmet with both head and face protection

Optional: Video camera and monitor

Safety Precautions

This demonstration should be performed only by experienced skaters. When performing the demonstration, be sure that a helmet is worn that will protect both the head and face of the demonstrator. Clear the demonstration area of obstacles that could contribute to injury if the demonstrator were to fall. This demonstration is best done in an empty hallway where obstacles would be a minimum.

Demonstration

While standing still on your skates, throw a small round plush toy up into the air and catch it as it comes back down to you. Ask students to explain the motion of the toy, being sure to discuss the context of the toy having a motion along the vertical axis. Next, use your skates to glide through the room at a constant velocity and have students discuss your motion as it pertains to the horizontal axis. Now, combine the two motions by moving about the room on your skates at constant velocity, repeatedly throwing the toy up into the air and catching it in your hand. Discuss with students the motion of the toy, and ask them to describe how the motion relates to the two axes of motion, which you have already illustrated. Ask students: Why does the ball fall back into your hand if you do not aim the ball forward as you toss the ball? As an option, you can use a ball-and-paddle toy. Discuss the various ideas and hypotheses students propose.

Reference Special thanks to Professor Eugenia Etkina, Graduate School of Education, Rutgers University, New Brunswick, NJ, for sharing this idea.

A commercial projectile cart demonstrator is used to illustrate the independence between horizontal and vertical motions of an object.

Application

Projectile motion • Two-dimensional motion

Parabolic trajectory • Frame of reference

Theory

The horizontal component of two-dimensional motion is independent of the vertical component. This demonstration employs a commercial projectile cart demonstrator to illustrate this concept. The cart and its cargo (a ball) is pushed across a table with a horizontal velocity. Midway along its path, the ball is shot vertically upwards using a built-in, spring-loaded, automatically-triggered mechanism. Since both the cart and ball are traveling at the same velocity, they continue traveling while remaining vertically aligned. The ball follows a parabolic path, rising upwards and then down—falling into the cup from which it originated.

Materials

Baseball or similar object

Projectile motion cart and ball*

Thread, 1 m

Optional: Ball and paddle toy

Note: Pasco® and Cenco® both produce Projectile Motion Carts. A homemade version may also be fashioned from the following materials:

Battery-operated car, slow-moving

Cup, 6-oz

Hot glue

Toy, spring-popping

Safety Precautions

Although this demonstration does not pose any obvious risks, always follow laboratory safety rules. Wear safety glasses when performing this demonstration.

Preparation

Secure a 1 m long thread to the release pin of the projectile motion cart.

As an option, a projectile motion cart may be created by using hot glue to attach a clear plastic cup to the top of a slow-moving, battery-operated car. A spring-popping toy is used in place of the automatically launched ball.

Demonstration

Walk around the room repetitively throwing a baseball up into the air and catching it in your hand. Ask students: Why does the ball fall back into your hand if you do not aim the ball forward as you toss the ball? As an option, you can use a ball and paddle toy. Discuss the various ideas and hypotheses students propose.

Compress the spring of a projectile motion cart, insert the release pin and then insert the ball. While holding onto the thread secured to the pin, gently push the cart away from your hand. Once the thread becomes taut, the pin will be pulled from the cart, causing the ball to be ejected from the muzzle of the launcher. The ball will then follow a parabolic path and land back into the muzzle of the launcher. Discuss the independence between horizontal and vertical motion of the ball.

Optional Demonstration

Compress the spring-popper toy and determine the time it takes for it to "pop." Activate the battery-powered toy car and let it travel across the demonstration table. Ask students what they think will happen if the popper toy is placed into the plastic cup and allowed to pop as the car is moving. After discussing the various ideas, reposition the car at one end of the demonstration table, place the compressed toy into the cup, and activate the car.

The popper toy should pop upwards and then fall back into the moving cup. Discuss the independence between horizontal and vertical motion of the ball.

Note: Consider mounting a mini wireless video camera to the car with its lens directed upward to show that the position of the ball is above the car during its entire flight.

Reference

Sutton, Richard M. "M-99: Relativity Car," *Demonstration Experiments in Physics;* McGraw-Hill: New York, 1938; p 46.

Chapter 4

Forces and Newton's Laws of Motion

What Is Inertia?

The mass of an object is demonstrated in terms of its inertia.

Application

Inertia • Mass • Inertial balance • Newton's First Law

Theory

The mass of an object is the quantity of matter in that object. It may be determined by measuring the motion of the object, which is affected by an object's mass. The most common method of measuring mass is using a two-pan balance. This method compares the unknown mass of an object to the known mass of an object. The known mass is called a standard. Another method of determining the mass of an object is to employ an inertial balance to measure the object's resistance to a change in motion. The amount of mass an object has is directly proportional to the object's inertia. An inertial balance works by comparing the time it takes a spring-like arm of the balance to swing the object back and forth a certain number of times. The arm of an inertial balance changes position more slowly when a mass is added to the end of the inertial balance. Inertial balances will work in a weightless environment such as an orbiting spacecraft unlike a two-pan balance, which requires a minimum amount of gravity to operate.

Materials

C-clamp, large

Inertia balance

Spring clamps (or C-clamps), 3

Wall clock with second hand or stopwatch

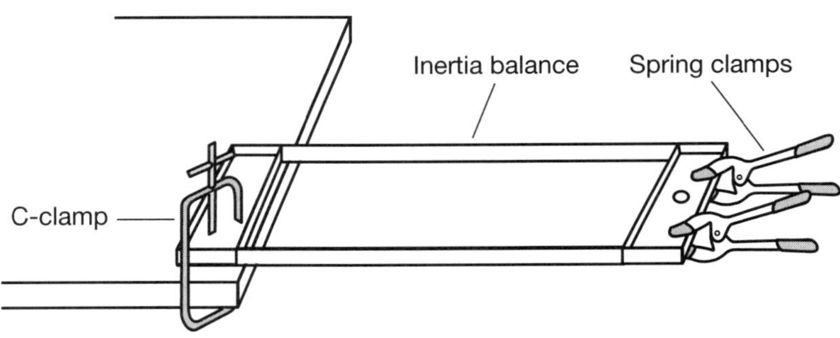

Safety Precautions

Although this demonstration does not pose any obvious risks, always follow laboratory safety rules.

Demonstration

Use a large C-clamp to secure the inertia balance to the top edge of a sturdy table. The opposite end of the device must be free to swing parallel to the floor.

Pull the end of the apparatus sideways about 10-cm and allow to swing back and forth. Determine the time it takes for 20 swings to pass. Record this value. The arm should swing smoothly back and forth for at least 15 s.

Discuss with students what they think will happen if a mass, in the form of a spring clamp, is attached to the end of the apparatus. Attach one clamp to the end of the apparatus and then pull the arm sideways 10 cm and allow to swing back and forth. Determine the time it takes for 20 swings to pass. Compare this value to the apparatus without the added mass. Add a second and then third clamp and repeat the experiment. The time taken for 20 swings should increase with the addition of more mass.

The demonstration could be followed up by a graphing exercise. Plot period as a function of mass and then use the graph to identify the mass of an unknown.

Reference

Gartrell Jr., Jack E. *Methods of Motion — An Introduction to Mechanics;* National Science Teachers Association: Washington, DC, 1989; p 21.

Newton, "the Magician" — His First Law

A classic magician's "trick" is performed and explained using Newton's First Law of Motion.

Applications | Newton's First Law • Inertia

Theory | Newton's First Law of Motion, sometimes called the Law of Inertia, states that an object at rest will remain at rest and an object in motion will remain in motion at a constant velocity unless an unbalanced force acts upon it. Examples of this law can be seen every day. A passenger standing in a bus or subway that is pulling away from a stop will experience the vehicle floor accelerating their feet but not their upper body. To prevent from falling backwards, passengers must hold on to a handle or strap to help the upper part of their body accelerate forward at the same rate as their feet.

Materials | Ceramic dinner plate (Corel® plates work well)

Cloth, smooth and seamless, 1 m × 1 m (satin works well)

Glass, drinking

Serving spoons, large metal, 2 (forks and knives are not recommended)

Salt and pepper shakers

Table surface, smooth and sturdy

Vase with heavy base and a flower

Alternative Materials | Sheet of paper, 8½″ × 11″

Soft drink bottles, glass, 2

Note: Objects must be smooth on their bottoms. Tall top-heavy objects such as wine glasses are not recommended. Bottom heavy objects, such as drinking glasses with thick glass bottoms, are desirable. To ensure stability, fill containers such as glasses or vases with some water. Forks and knives are not recommended as they pose a hazard if they become airborne.

Safety Precautions | Wear safety glasses while performing this demonstration. Follow all laboratory safety rules.

Demonstration | Arrange the place setting using the satin as a tablecloth. Do not permit the cloth to hang over the audience edge of the table. Stand back about half a meter from the demonstration table. Grab the ends of the cloth and gently pull until taut. With one quick and continuous motion directed towards the floor pull the cloth from *beneath* the setting to prevent the cloth from rising above the table surface. The setting should remain undisturbed. Ask students if they can explain why the setting does not move with the cloth.

Place a wide-mouth glass soda bottle on a sturdy surface. Place one edge of a smooth sheet of paper (8½″ × 11″) over the opening of the bottle. Now place a second, but inverted, bottle on top of the paper and first bottle. *Note:* Two Classic Coke® bottles work well. The demonstration involves quickly pulling the sheet of paper from between the bottles. The bottles should remain standing—one on top of the other. Ask students if they can explain why the top bottle does not fall over.

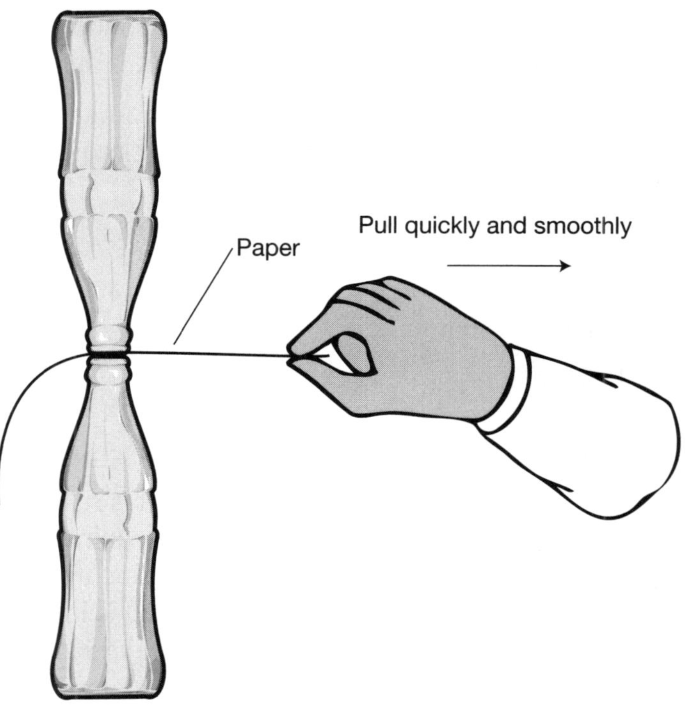

Inertia — Keeps Things Put!

The inertia of an object will prevent it from moving.

Application | Inertia • Newton's First Law • Jerk • Tension • Tensile strength

Theory | Inertia is the property of matter that manifests itself to resist any changes in its motion. That is, a body in motion will remain in motion and a body at rest will remain at rest unless an external force acts upon it. The force, F, must be great enough to overcome the body's inertia to change motion. Mass, m, is the measurement of an object's inertia. In this demonstration a bowling ball is lifted by a string. When the string is pulled with a constant velocity (or with a low rate of acceleration), the ball is lifted without the string breaking. If the system is accelerated, then the string will break if the rate of acceleration is sufficient.

When the system is in static equilibrium (stationary or moving at a constant velocity), the tension, T, in the string is equal to the weight, w, of the ball:

$$\Sigma F = m(0) = T - w$$
$$T = w$$

When the ball is accelerated, the tension in the string increases in direct response to the acceleration, a:

$$\Sigma F = ma = T - w$$
$$T = w + ma$$

If the tension in the string exceeds its tensile strength limit, the string will break.

The string used in this demonstration is relatively strong: $1/16''$ nylon. Consequently, a rather large rate of acceleration is required to break the string. The large rate of acceleration may be achieved by pulling up on the ball quickly and increasing (i.e.; changing) the upward force as the ball is lifted. Subsequently, the acceleration increases over time. A change in acceleration is called a jerk, jolt, surge or lurch. A jerk is the derivative of acceleration, or the rate of change of acceleration with respect to time. Engineers must consider the jerk experienced by passengers in the design of roller coasters, subways and other vehicles to prevent injury (example: whiplash) and motion sickness.

Materials | Adhesive, construction or plumber's epoxy

Bowling ball, 12–16 lb

Drill

Drill bit, $7/16''$

Mason's line, nylon, $1/6''$

Screw eye, $3'' \times 1/2''$

String, thin cotton (the type used in bakeries for tying packages)

Wooden dowel, $3/4'' \times 6''$ long

Although this demonstration does not pose any obvious risks, always follow laboratory safety rules. Take care not to drop the bowling ball on your feet.

Preparation

Use a $^{7}/_{16}''$ drill bit to drill a 3″ deep hole into the bowling ball. Squeeze a small amount of construction adhesive or plumber's epoxy into the hole and insert the screw eye. Use a pair of pliers or a large screwdriver for leverage, if necessary.

Tie a 40-cm piece of cotton string between the screw eye and a piece of ¾″ dowel. The dowel will serve as a handle that is used to lift the bowling ball.

Demonstration

Position the bowling ball on the floor. Ask students if they think the bowling ball can be lifted using the thin cotton string. Slowly lift the ball using the handle and string to verify that the string can support the mass of the bowling ball. Discuss the free body diagram of the system. Ask students what will happen if the bowling ball were pulled upwards suddenly while suspended from the string. After observing what happens, ask students to explain why the string breaks in this case. This result occurs if the action begins with the ball resting on a surface or suspended in air.

References

Heald, Mark A.; Caplan, George M. "Which String Breaks?"; *The Physics Teacher;* American Association of Physics Teachers: 1996; Vol. 34, Nov. pp 504–507.

Sandin, T. R. "The Jerk"; *The Physics Teacher;* American Association of Physics Teachers, 1990; Vol. 28, pp 36–40.

Pull It!

Measurement of tension is explored using various configurations of spring scales.

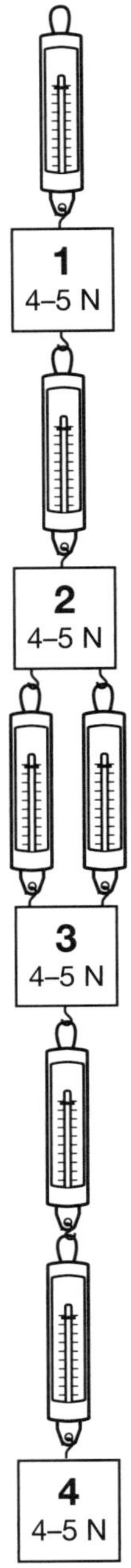

Application | Tension • Free body diagrams • Equilibrium • Newton's Second Law

Theory | Tension is a force or stress that pulls or stretches an object. When an object is suspended from a string connected to a rigid body (such as a hook mounted to a ceiling) tension exists in the string due to the downward pull of the weight of the object. Such a system is also said to be in equilibrium since the downward pull of the object acting on the string is counterbalanced by the upward pull of the rigid body.

A free body diagram (FBD) is a simplified representation of an object (the body) in a problem showing all the forces and their directions acting on the object. The body is considered "free" since it is shown without its surrounding environment—eliminating unnecessary information. In this demonstration, four blocks are connected in series and suspended from spring scales in various configurations. Students analyze the system as blocks are added one after another—each time summarizing their analysis by sketching an FBD of the system. The demonstration also focuses on how tension can be divided, as illustrated in the two scales between blocks 2 and 3. Tension is also constant between two points as illustrated in the two scales connected between blocks 3 and 4.

Materials | C-clamp

Cross bar

Hook collar clamp or suspension hook clamp

Right angle clamp

Ring stand, large

Screw eyes, ⅜″, 9

Spring scales, 20 N, 6

Wood blocks, 4″ × 4″ × 4″, 4 [each weighing 4–5 N (400–500 g)]

Safety Precautions | Although this demonstration does not pose any obvious risks, always follow laboratory safety rules.

Preparation | Attach screw eyes to the blocks of wood, in order to suspend the blocks as illustrated in the diagram.

Secure the ring stand to the demonstration table using a C-clamp. Use a right angle clamp to secure a crossbar to the top of the stand. Attach a hook collar clamp or suspension hook clamp to the end of the crossbar, from which the demonstration apparatus will be suspended.

<table>
<tr><td>Demonstration</td><td>

Suspend the first block from a spring scale suspended from the hook collar clamp. Invite a student to read the spring scale to the class. Have students sketch a free body diagram of the system in their notebooks. Ask students to predict what the reading would be if a second block were added to the system, but suspended from block 1 via a spring scale. Ask students to predict the readings on both scales. After discussing the results, have students sketch an FBD of this system in their notebooks. Next, ask students to predict what the reading would be if a third block were added to the system, but suspended from block 2 via two spring scales connected in parallel. Again, have students summarize this second system by drawing an FBD in their notebooks. Next, add a fourth block suspended by two scales connected in series. Direct students' attention to the fact that the tension is divided between the two scales located between blocks 2 and 3, but found to be constant between blocks 3 and 4.

As an alternative demonstration, have the entire apparatus set up at the front of the room and invite students to predict the readings on each spring scale. Later have them approach the apparatus and verify their predictions.

</td></tr>
</table>

Push It!

The forces acting on a cart of books are compared with the forces on a cart at rest on a horizontal surface, first as the carts move across the surface and later as they move up an inclined plane.

Application

Vector components • Free body diagrams • Normal force • Newton's Second Law

Theory

To change an object's state of motion—that is, to move, stop or slow it down—an adequate force must be applied to the object. In this demonstration the forces acting on a rolling cart of books as it is moved across the floor are explored in terms of magnitude, direction and resulting motion. A Free Body Diagram (FBD) will be sketched for each case in the demonstration. An FBD is a simplified representation of an object (the body) in a problem showing all the forces and their directions acting on the object. The body is considered "free" since the body is shown without its surrounding environment—eliminating unnecessary information. In this demonstration, the cart represents the body in question.

Initially, the cart is at rest. The only forces acting on the cart are the downward pull of gravity, w, and the normal force, N, which acts in response to the downward force. In the FBD, the lengths of the arrows representing N and w are identical to emphasize that their magnitudes are equal.

In the second case, a student pushes the cart across the floor. Since the cart has a very large mass, it will be apparent to all observers that the student will have to apply a greater force to initiate motion than he or she will need to apply to keep the cart in motion. The initial force applied must overcome the static friction, f_s, acting on the cart, which is greater than the kinetic friction, f_k, experienced once the cart is set in motion. Note that the applied force vector, F, is greater than f_s. Once the card is set in motion, two FBDs are initially possible—one illustrating the cart accelerating in which $F > f_k$ and a second illustrating the cart traveling at a constant speed in which $F = f_k$. The equations that summarize the forces in the case of the cart accelerating are: $\Sigma F_x = ma = F - f_k$ and $\Sigma F_y = 0 = N - w$.

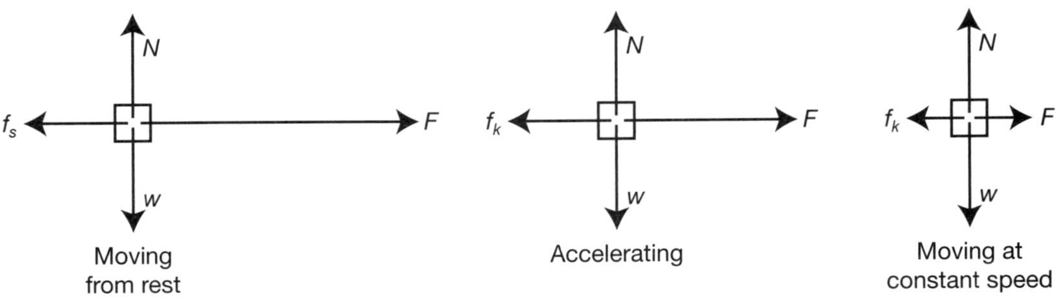

If the student were to decrease the applied force such that $F < f_k$, the cart will begin to decelerate. If $F = 0$, the rate of deceleration will be maximum and solely dependent upon f_k. That is, $\Sigma F_x = ma = -f_k$

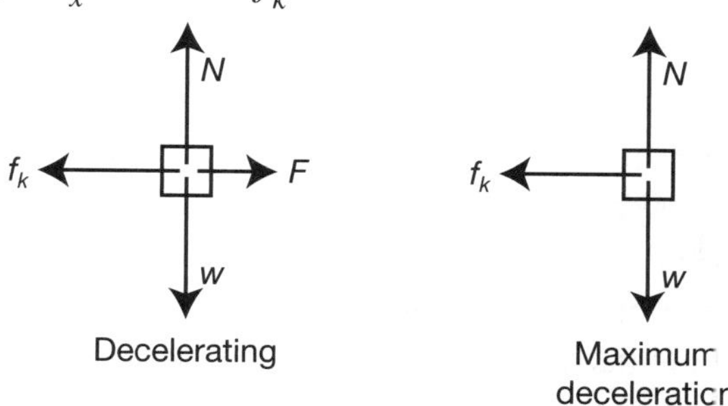

When the student pushes the cart up a wheelchair ramp it becomes very obvious that a much greater force must be applied than in any of the previous examples. The reason for this is that F is used both to move the cart across the plane as well as to lift the cart against the force of gravity as it gains elevation traveling up the ramp. In this situation, the axes of the FBD are defined with respect to the ramp, rather than with respect to the floor in order to simplify the analysis. The force vectors must all be in terms of x and y. The equations that summarize the forces in this case are: $\Sigma F_x = ma = F - f_k - w\sin\theta$ and $\Sigma F_y = 0 = N - w\cos\theta$.

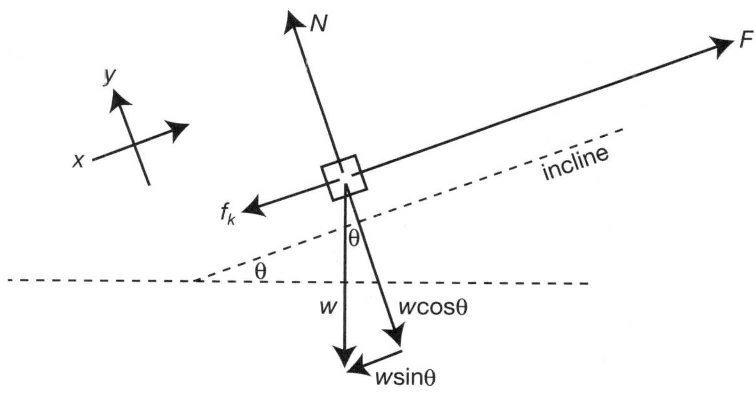

Materials

Metal AV cart on wheels completely filled with books, (approx. 60 volumes of an encyclopedia or textbooks)

Wheelchair ramp

Optional: Bathroom scale

Safety Precautions

Although this demonstration does not pose any obvious risks, always follow laboratory safety rules.

Direct students' attention to an AV cart filled with books. Ask students what sort of forces act on the cart when the cart is at rest. Review all the forces acting on the object including weight and the normal force. Draw a Free Body Diagram of the forces acting on the cart.

Ask students what must be done in order to move the cart. Most students will respond by saying that a force is required to change the object's state of motion. Invite one student, preferably a strong student, to push the cart gently across the room at a constant rate. Have the student describe the force he or she exerts to keep the object in motion. To measure the force exerted, consider placing a bathroom scale between the hands of the student and the cart. Have the same student gently push the cart across the floor and then release the cart, so that it continues moving across the floor on its own until it comes to rest. Note how far the cart traveled. Ask students what the travel distance of the cart depends on. Review all the forces acting on the object including weight, the push force, the normal force and friction. Draw a Free Body Diagram of the forces acting on the cart.

To verify that the distance depends upon the mass of the cart and books, remove half of the books and have the same student push the cart across the floor and then release the cart, so that it continues moving across the floor on its own until it comes to rest. Compare the distances that the cart traveled, filled and half-filled. The half-filled cart should travel further. Discuss the relationship of the distance traveled after release with respect to the rolling friction of the wheels.

Finally, ask students how much force would be needed to push the same cart on an inclined plane. Have the same student push the book-filled cart gently up a wheelchair incline. Have the student describe the force he or she exerts to keep the object in motion. Review all the forces acting on the object including the component and direction of the weight acting on the cart, the push force, the normal force and friction. Draw a Free Body Diagram of the forces acting on the cart.

Move It!

Explore the angles and magnitude of the components of vectors with respect to the original vector.

Application	Vector components • Statics • Equilibrium
Theory	In this demonstration, a rope is pulled by two strong students (S_1 and S_2) while a third, weaker, student (W) pushes down in the middle of the tight rope. No matter how hard the two strong students try, they are unable to keep the rope taut. The three forces acting on the rope create a system at static equilibrium. For the rope to be kept perfectly taut, S_1 and S_2 must be kept at 90° with respect to W. As S_1 and S_2 approach 90° their magnitude approaches ∞. At 89°, S_1 and S_2 are 29 N each when W applies a mere 1 N of force.

Materials

Carpenter's level

Mass, 1 kg

Ring stands, tall, 2

Rope, ¼″, 3–4 m length

String, 2 m

String, 50 cm

Safety Precautions

Although this demonstration does not pose any obvious risks, always follow laboratory safety rules.

Preparation

Set two ring stands in the middle of the demonstration area on top of two desks. Tie a 2 m long string from one ring stand to the other. Adjust the height of the string to mid-chest level of the demonstration assistants, S_1 and S_2. The string must be horizontal, as it will serve as the reference line for the demonstration.

Demonstration

Ask the class to identify the two strongest individuals in the classroom and invite them to join in the demonstration. Have these two students, S_1 and S_2, pull on opposite ends of the rope in such a way to keep it aligned with the reference string secured to the ring stands. Ask the volunteers to try to keep the rope taut so that the rope is always horizontal and never drooping. They should not engage in a tug of war.

Invite another student to assist in the demonstration. Ask the class if they think that this student, W, can overcome the strength of students S_1 and S_2. Discuss the various ideas that the class puts forth. Tie a 50-cm string to the center of the rope and ask student W to pull down on the string while the stronger students pull with all their might. Have students focus their attention on the rope with respect to the reference line. Inevitably the third student, W, is deemed to be stronger because she has physics on her side. Finally, review why it is impossible to keep the rope horizontal in this demonstration. As an option, a 1-kg mass can be suspended from the rope in place of W.

A Darting Demo of the Second Law

The effect of mass on acceleration is illustrated using two dart guns.

Application | Newton's Second Law • Hooke's Law

Theory | In this demonstration, two darts are launched from separate but identical dart guns. One dart is made more massive by permanently attaching a steel marble. When the darts are fired toward the ground simultaneously, the lighter dart reaches the ground first. According to Newton's Second Law, the greater the mass of an object, the slower the acceleration if the same force is applied. In this case, the force is identical, so the heavier dart moves out of the barrel at a slower rate. The spring of the gun provides the force: $F = kx$, where k is the spring constant and x equals the distance the spring is compressed.

Materials | Adhesive, construction

Ball, marble or steel, ¾″

Dart guns with rubber suction cup darts, 2 identical

Note: Not all dart guns are alike. Although the safety of modern dart guns has improved, many of these guns are not adaptable to this demonstration. The type of gun needed has a dart with a rigid shaft that locks into the spring mechanism of the gun. Foam and rubber dart guns do not work in this demonstration.

Safety Precautions | Never aim the dart gun at anyone. Darts have been known to spontaneously launch without warning. Wear safety glasses when performing this demonstration. Always follow all laboratory safety rules.

Preparation | Use construction adhesive to glue a marble or steel ball to the suction cup end of one rubber dart.

Demonstration | This demonstration is best done if standing on the edge of the demonstration table.

Load one dart gun with a regular dart and the other with the dart that has the marble attached. Point the guns downward, aiming for the floor. Be sure that all students will be able to see the path of the darts as they are fired. Ask students: Which dart will hit the floor first if they are launched simultaneously? Many students will say that they should reach the ground at the same time, since "all objects fall at the same rate." Launching the darts will result in the marble-free dart reaching the ground first. Discuss the paradox.

Note: The motion of the objects in this demonstration is rather fast resulting in a limited time for observation. Consider videotaping the demonstration and reviewing the tape in slow motion or frame by frame.

Reference | Carpenter Jr., D. Rae; Minnix, Richard B. "Dart Guns" in *The Dick and Rae Physics Demo Notebook*. Dick and Rae: Lexington, Virginia, 1993; p M-288.

Law of Interaction — *Newton's Third One*

A small air rocket is launched across the room to illustrate Newton's Third Law.

Application

Newton's Third Law • Action–reaction force pair • Space flight

Theory

One of Isaac Newton's most important discoveries, his Third Law, states that if a force acts on a body, then an equal and opposite force must act upon another body. The single most important concept of his law is that action and reaction forces always act upon *different* objects. That is, when one body exerts a force on another, the second body exerts an equal but opposite force on the first body. The two forces are balanced. For example, a person exerts a force on a wall and the wall exerts an equal but opposite force on the person—two forces arising from and acting on different objects. In contrast, the often cited example of a book lying on a table is not an example of Newton's Third Law when erroneously explained as the weight of the book as the action and the upward Normal force as the reaction, since both forces act on the same object. For this reason, a Free Body Diagram does NOT illustrate Newton's Third Law since the force acts only on one body. The correct interpretation of the book example would identify the action as the force of attraction (i.e., gravity) of the book to the Earth and the reaction as the force of attraction of the Earth to the book.

The single most important example of Newton's Third Law illustrates the principle of rocket flight: As the rocket pushes out, the exhaust pushes the rocket forward. In this demonstration, a balloon attached to a guide wire is launched across the room. The stretched balloon exerts a backward force—the action force—by forcing the air out of the rocket. The air from the balloon exerts a forward force on the balloon—the reaction force. Although the surrounding air may have some negative or positive effect on the movement of the balloon in this demonstration, these effects have no relation to the fundamentals of rocket flight. A rocket would behave similarly in the vacuum of space.

Newton's Third Law has always been proven to be easily misunderstood. In response to rocket pioneer Robert Goddard's proposal to launch a rocket into space, there was a *New York Times* editorial on January 20, 1920 mocking him and questioning his hypothesis that rockets could travel in the vacuum of space. The paper's poor understanding of action–reaction pairs is underlined by their statement:

> "... Professor Goddard... does not know the relation of action to reaction, and of the need to have something better than a vacuum against which to react—to say that would be absurd. Of course he only seems to lack the knowledge ladled out daily in high schools."

A day after the Apollo 11 launch, in July 1969, the *New York Times* published "A Correction" and retracted its 1920 editorial—in effect acknowledging Goddard's mastery of high school physics.

<table>
<tr><td align="right">Materials</td><td>

Oblong balloon

Screw eyes, 2, or heavy ring stands, 2, and C-clamps, 2

Soda straw

String, 5 m

Tape
</td></tr>
<tr><td align="right">Safety
Precautions</td><td>

Although this demonstration does not pose any obvious risks, always follow laboratory safety rules.
</td></tr>
<tr><td align="right">Preparation</td><td>

Thread a string through a soda straw and suspend the string across the length of the classroom. The string can be connected to two screw eyes screwed into the walls/cabinets of the classroom or strung between two heavy ring stands positioned at opposite ends of the room.
</td></tr>
<tr><td align="right">Demonstration</td><td>

Fill an oblong balloon with air. Keeping the balloon sealed by pinching the nozzle, tape the balloon parallel to the straw using two pieces of tape. Release the balloon. As the air leaves the balloon, it should travel down the string to the opposite side of the room. Ask students to identify the action and reaction forces of this system. You can use this demonstration as a springboard to discuss other examples of action–reaction force pairs.
</td></tr>
<tr><td align="right">Reference</td><td>

"Topics of the Times: A Severe Strain on Credulity"; *New York Times;* January 13, 1920; p 12. A correcting, unsigned editorial, *New York Times;* July 17, 1969; p 43.
</td></tr>
</table>

A student standing on a skateboard jumps off but does not move—illustrating that forces which cause motion act in pairs.

Application	Newton's Third Law • Action–reaction force pairs
Theory	One of the most cited examples of Newton's Third Law is a person attempting to step off of an unsecured boat onto a dock. The action is described as the person applying a force to the boat, which results in the movement of the boat away from the person. Consequently the person falls into the water. The example can be described in terms of Newton's three laws and action–reaction force pairs. According to the Third Law, the person applies a force to the boat causing the boat to accelerate in one direction, while the boat applies an equal and opposite force that causes the person to accelerate in the opposite direction. It is important to note that action and reaction forces always act upon *different* objects but occur during a single event. According to Newton's Second Law, the relative acceleration depends on the mass of the objects. If the boat is relatively small, such as a canoe, it will accelerate at a much faster rate than the person due to the large difference in mass. In this demonstration, a student attempts to perform a long jump while standing on a skateboard. The principles in this case are similar to that in the boat example. The person pushes backward on the skateboard causing it to accelerate in one direction, while the skateboard pushes with an equal but opposite force on the person causing the person to accelerate in the opposite direction. Because of the large difference in mass, the skateboard is expected to move very fast and far making the displacement of the person hardly noticeable. Once the person hits the floor, the friction between his or her shoes and the floor surface also contributes to the deceleration of the individual.

Another example of action–reaction pairs includes a person standing on the ground. The Earth exerts a downward directed attractive force on the person (action), while the person exerts an upward directed attractive force on the Earth (reaction). If the person were to jump upward, he would exert a downward force on the Earth while the Earth exerts an equal but opposite force on the person resulting in the two objects accelerating away from one another. Many students will find this example harder to accept since the movement of the Earth is not noticeable. To illustrate this concept, have a person throw a medicine ball away from her body while sitting on a skateboard or standing on roller skates, as outlined in the demonstration *"Rollerball,"* page 47. In this case, the acceleration of both objects will be obvious.

Materials	Football helmet
	Large skateboard

When performing this demonstration, be sure that a helmet is worn that will protect both the head and face of the demonstrator. Clear the demonstration area of obstacles that could contribute to injury if the demonstrator were to fall. This demonstration is best done in an empty hallway where there are few obstacles. Leave plenty of room behind the volunteer for the skateboard to move along. Be prepared for the skateboard to move very fast and far. Stand to the side and ahead of the volunteer and be prepared to prevent the volunteer from tipping forward as he or she attempts the feat.

Demonstration

Invite a student to assist in performing the demonstration. It is best to select a student who has athletic coordination. A student with experience doing the standing long jump is ideal. Have the student wear a football helmet that protects the head and face. Position the skateboard in the middle of the demonstration area. Instruct all spectators to stand along the side of this area—in front of the student volunteer. Ask the student to perform a practice long jump on the floor surface—taking note of the maximum displacement. Next, instruct the student to stand on the skateboard. Ask him or her to do a standing long jump from the skateboard. If the long jump is performed correctly, it will result in the skateboard being kicked forcefully backward, with the student hardly moving forward at all. Expect the skateboard to travel very fast and far away from the volunteer. Ask students to discuss why this latest attempt resulted in such a small displacement. Review the action and reaction forces present in this system and emphasize that they act on different objects but occur at the same time. Discuss other examples of action–reaction pairs.

The difference between weight and apparent weight is explained through the concept of a Normal force.

Application   Apparent weight • Normal force • Newton's Third Law

Theory   An object, such as a book resting on a table, is subjected to two forces: the force of gravity pulling it downward and the Normal force of the surface of the table counteracting gravity. The Normal force arises from the repulsive forces of the atoms of the surfaces of the book and table as well as the bonding forces holding the atoms of the table together. The atoms can be thought of as held together by springs to help illustrate the concept. Placing a heavy object, such as a concrete block, on a thin board supported at either end would cause the board to bend. This shows how the Normal force (i.e., the combination of repulsive and bonding forces) counteracts the force of the block pushing on the board. The greater the push, the greater the upward Normal force. The Normal force always acts perpendicular to the surface. A Normal force is not limited to counteracting the effects of gravity. A horizontal force applied to a wall is countered by a Normal force in the opposite direction.

In this demonstration a person stands on a bathroom scale. The Normal force, N, would be equal in magnitude to the weight, w, of the person since no other forces are involved, as illustrated in Figure 1.

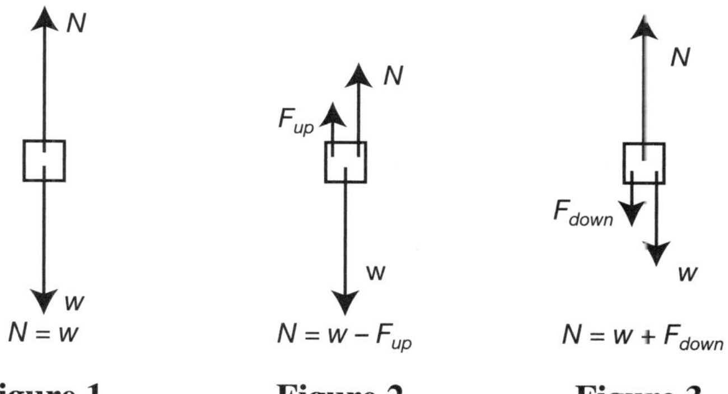

Figure 1.   **Figure 2.**   **Figure 3.**

Although it may be tempting to describe the situation in terms of Newton's Third Law, students should be reminded that the weight of the person on a scale is not paired with the Normal force provided by the scale in terms of action–reaction force pairs. Neither force can be considered a reaction force of the other. It is important to note that action and reaction forces always act upon different objects but occur during a single event. In the case of gravity, the Earth exerts a downward directed attractive force on the person (action), while the person exerts an upward directed attractive force on the Earth (reaction).

Remind students that it's common for people to stand on a bathroom scale and lean against the towel rack (a downward push) or sink, resulting in a lower reading on the scale. By pushing down on the towel rack, the rack applies an upward force, F_{up}, that, in addition to the Normal force, counteracts the downward force of gravity, as illustrated in Figure 2. In this case, the scale is not measuring the true weight of the person, but is instead showing the apparent weight of the person since a third force, F_{up}, is involved. The apparent weight is the Normal force.

If the person pushes upward on the towel rack, the scale will show a reading that is greater than the true weight of the individual, as illustrated in Figure 3.

Having completed this demonstration, one may want to start referring to a bathroom scale by a more descriptive name—a Normal Force Meter.

Materials

Bathroom scales, 2

Doorway

Safety Precautions

Although this demonstration does not pose any obvious risks, always follow laboratory safety rules.

Demonstration

Position yourself in the middle of a door frame. Stand on a bathroom scale and invite a student to read the scale. Ask students: What does the reading on the scale measure? Does it measure weight or something else? Most likely, students will say that it measures weight.

While standing on the scale, pull yourself up using the top of a door frame. Ask a student to read the scale in this situation. Your weight will seem to be lower. Ask your students why this is the case. Explain that the measurement is your *apparent weight,* because there is another force, F_{up}, acting on your body, in addition to the pull of gravity.

Hold a second bathroom scale in your hands above your head. While standing on one scale, position the second scale against the bottom surface of the top of the door frame. While pushing upwards, ask a student to read the scale in this situation. You will appear to have gained weight. Ask your students why this is the case. The difference between the readings on the two scales is equal to your actual weight.

Now, have two students stand on separate bathroom scales and face each other with their hands extended. Have one push down on the other's palms. Note what happens to the readings on the scale. The apparent weight of the student pushing downward will decrease, while the apparent weight of the other will increase by the same amount.

Finally, ask students to explain what the bathroom scale actually measures in all of these cases. The scale measures the Normal force exerted on the surfaces of the floor and the door frame.

Reference

Thanks to William Koenig of Pascack Valley High School, Hillsdale, NJ for his input in developing this demonstration.

The friction force of two tire treads is compared.

Application | Friction • Coefficient of friction • Centripetal force

Theory | Automobile tires are designed to provide friction and to redirect water from between the tire and the road surface. The rubber surface, when in contact with the road, provides friction to keep the tire from slipping as it rolls along. The grooves in the tire provide channels for water and other debris to be directed away from the tire, allowing constant contact between the rubber tire and the road. If a fast moving car encounters a large pool of water, the grooves of the tire may be insufficient to channel the water away, causing the car to hydroplane uncontrollably across the surface of the water. The tires of a car must be replaced on a regular basis because the depth, and therefore the effectiveness of the grooves, will decrease over time. Laws require that automobile tire treads have a minimum depth to be roadworthy. Ironically, on dry surfaces free of debris, a bald tire may provide greater friction than a tire with its treads intact because the bald tire provides a larger contact area between the rubber tire and the road surface. For this reason the tires used in automobile racing are very wide and tread-free. The race, however, must take place on a dry track. In the event of a shower, the race is interrupted until the shower passes. The track is then dried and cleaned of debris using air blowers before the race is allowed to resume.

Materials | Automobile tires, 2, one with a good tread and one bald (available from a tire center or recycling center)

Adhesive, construction

Drywall screws, 1½″, 1 lb

Nylon rope, ¼″, 6 m

Reciprocating saw or hacksaw

Screw eyes, ½″, 2

Solid concrete blocks, 4″ × 8″ × 16″, 3–4

Wire cutters

Wooden boards, 2″ × 8″ × 18″, 2

Safety Precautions | Review manufacturers' recommendations before working with power tools. Wear safety glasses when operating power tools. Although this demonstration does not pose any obvious risks, always follow laboratory safety rules.

Preparation

Secure a ½″ screw eye into one end of each 2″ × 8″ board.

Use a reciprocating saw or hacksaw to cut a 6″ × 18″ piece of tire tread from each tire. Use wire cutters to trim any exposed wire from the steel belts that may be protruding from the tire. Each piece must lay flat. If they do not, you can make small horizontal cuts into the side of the tread. Apply a liberal amount of construction adhesive to the underside of one tire tread and secure it to one board using drywall screws. Be sure that the screw heads are imbedded into the rubber so that they will not scratch the surface on which the tread will be dragged. It is important that the tread surface is as level as possible—adjust the depth of the screws as necessary. Secure the second tread in the same manner. Allow the adhesive to set overnight.

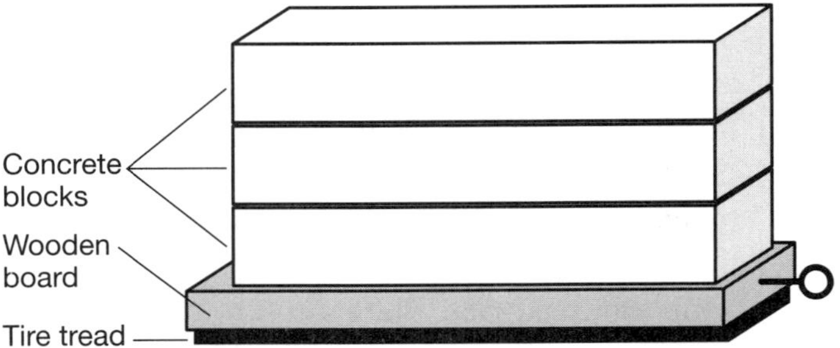

Demonstration

The demonstration should be conducted on smooth and flat pavement that is dry in one area and has a puddle of water in another area. The puddle should be at least ¼″ deep and 5 m long.

Show students the tread boards. Ask students to predict which tread would stop a car better? Most students will probably think that the good quality tread will work better. Invite one student to lift and inspect each tread to confirm that they are essentially the same except for the different treads. Load the tread boards with 3–4 concrete blocks. Ask the student to separately drag each tread board across the dry pavement and describe which is more difficult. The student will find that the bald tread was harder to drag across the pavement.

Have the same student repeat the experiment, but dragging each tread board through the puddle. The student will find that the good tread was harder to drag across the wet pavement. Ask students to explain the paradox.

Finally, ask students why racing cars use smooth tires when racing on a track.

The magician's tablecloth trick (Newton, "The Magician" — His First Law, p 54) is further analyzed.

Applications

Friction • Jerk • Newton's First Law • Inertia

Theory

In the demonstration, *Newton, "The Magician" — His First Law* (page 54), a table-cloth, m_2, was pulled from beneath a table setting, m_1, leaving the setting in place. The "trick" only works if the tablecloth is smooth. The low coefficient of friction between the setting and the tablecloth facilitates the smooth movement of the cloth.

The demonstration may be analyzed in terms of Newton's Second Law: $\Sigma F = ma$. When one imposes a low acceleration, the setting and tablecloth system move *together* with a rate of acceleration "a":

$$F_{applied} = (m_1 + m_2)\, a \quad \text{or} \quad a = \frac{F_{applied}}{(m_1 + m_2)}$$

At low rates of acceleration the setting moves with the tablecloth due to the static friction between the setting and the tablecloth: or $f = m_1\, a = \mu_s m_1 g$ or $a = \mu_s g$. If the acceleration of the system exceeds $\mu_s g$, then the setting cannot move with the tablecloth. Therefore, $a \geq \mu_s g$ is the condition for the dishes and tablecloth to move together. In terms of the force applied: $F_{applied} \leq (m_1 + m_2)\, \mu_s g$ for the setting and tablecloth to move together. If $a > \mu_s g$, the tablecloth will slide out leaving the dishes behind. In terms of the force applied: $F_{applied}$ must exceed $(m_1 + m_2)\, \mu_s g$ if the setting is to be left behind.

If the applied force drops below the critical value once the dishes are sliding, the trick may still work, since the coefficient of kinetic friction, μ_k is less than μ_s. (Once you get something sliding, it is easier to keep it moving.)

Achieving a sufficient rate of acceleration poses a challenge. The cloth must be quickly pulled (snapped) from beneath the setting in order for the trick to work. If the cloth is pulled slowly, the setting moves with the cloth. A large rate of acceleration may be achieved by increasing (changing) the applied force as the tablecloth is subsequently pulled, increasing the acceleration over time. A change in acceleration is called a jerk, jolt, surge or lurch. A jerk is the derivative of acceleration or the rate of change of acceleration with respect to time. Engineers must consider the jerk experienced by passengers in the design of roller coasters, subways and other vehicles to prevent injury (example: whiplash) and motion sickness.

Materials	Cloth, smooth and seamless (satin works well), 1 m × 1 m piece
	Dinner plate, ceramic (Corel® plates work well)
	Drinking glass
	Salt and pepper shakers
	Smooth and sturdy table surface
	Vase with a heavy base and a flower

Alternate Materials

Soft drink bottles, glass, 2

Sheet of paper, 8½″ × 11″

Note: Objects must be smooth on their bottoms. Tall, top-heavy objects such as wine glasses are not recommended. Bottom-heavy objects, such as drinking glasses with thick glass bottoms, are desirable. To ensure stability, fill containers such as glasses or vases with some water. Forks and knives are not recommended as they pose a hazard if they become airborne.

Safety Precautions

Wear safety glasses when performing this demonstration. Although this demonstration does not pose any obvious risks, always follow laboratory safety rules.

Demonstration

Arrange the place setting on a satin tablecloth. Do not permit the cloth to hang over the audience edge of the table. Remind students that they saw this demonstration before as presented in terms of Newton's First Law. Stand back about half a meter from the demonstration table. Grab the ends of the cloth and gently pull until taut. Slowly pull the cloth. The setting will move with the cloth when pulled slowly. Ask students why the place setting moves with the cloth this time. Next, with one quick and continuous motion directed towards the floor pull the cloth from beneath the setting. The setting should remain undisturbed. Ask students why the trick works when pulling quickly versus slowly. Emphasize that a jerk facilitates the ability to overcome the static friction.

References

Hudson, H. T. "There's More to It Than Inertia;" *The Physics Teacher;* 1985; vol 23, p 163.

Haber-Schaim, Uri; Dodge, John D. "There's More to It Than Inertia;" *The Physics Teacher;* 1991; vol 29, p 56.

Thanks to Dr. Mark Croft and Dr. Mohan Kalekar, Physics Department, Rutgers University, Piscataway, NJ for their input.

Compare the relative motion of two toy vehicles engaged in a tug of war with and without friction.

Application

Friction • Coefficient of friction • Vectors • Forces

Theory

The friction between two surfaces depends on two factors: the coefficient of friction between the surfaces and the normal force imposed by one surface onto the other. This relationship is summarized by: $f = \mu N$, where f represents the force of friction opposing motion, μ is the coefficient of friction and N is the normal force. In the case of an object resting on a surface, such as a toy car on a table, the Free Body Diagram may be represented as:

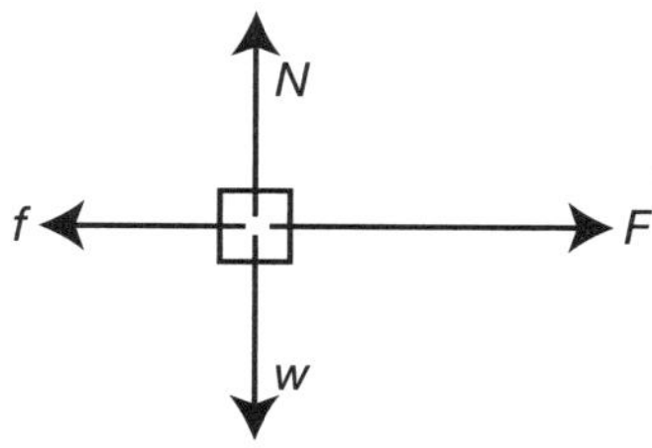

where w represents the downward force (weight) of the object acting on the surface. In this case $N = w$. In this demonstration two slow-moving toy vehicles are connected by a string and then set in motion in opposite directions to create a tug of war. Since the cars are identical, neither can overcome the other. Adding a mass to Car A increases its weight and therefore the Normal force acting on Car A increases, creating more friction. The greater friction enables Car A to defeat the other car. In a separate scenario, a piece of sandpaper is slipped under the tires of Car B. This increases its friction and causes Car B to overcome Car A.

Materials

Mass, 1 kg

Sandpaper, ⅜″ × 10″ sheets, taped end-to-end

String, 50 cm

Toy vehicles, identical battery-operated, constant-velocity (preferably, slow-moving), 2

Note: A cup hook glued to the back of the toy cars may be used if the vehicles do not have a rear bumper to which a string may be secured.

Safety Precautions

Although this demonstration does not pose any obvious risks, always follow laboratory safety rules.

Effect of Surface on Motion

Using the string, tie together the rear ends of two identical battery-operated, constant-velocity toy vehicles. Place the vehicles on a demonstration table in front of the students. Direct students' attention to the fact that the vehicles are identical and joined together. Ask students: What will happen if the motors of both vehicles are turned on? After a short discussion, turn on both vehicles and watch as the vehicles barely move back and forth in a tug of war that results in no overall change. Ask students: What will happen if one of the cars is placed on sandpaper while the other remains on the smooth surface of the demonstration table? Place the rotating wheels of one vehicle on the end of the strip of sandpaper, while the other vehicle remains on the bare surface of the demonstration table. Turn on both vehicles and observe how the vehicle on the rougher surface wins the contest.

Effect of Mass on Friction

Once again, place the vehicles on the bare demonstration table in front of the students. Ask students what will happen during a tug of war between two vehicles if one were heavier than the other, but no other differences exist? After a short discussion turn on both vehicles and position a 1 kg mass on the vehicle. Observe how the heavier vehicle wins the contest.

Chapter 5

Work, Energy, and Power

Work is Going My Way

Lifting and then carrying a heavy block clearly illustrates the concept of work.

Application | Work • Force • Acceleration • Energy

Theory

The amount of work done is equal to a force applied over a distance: $W = F \times d$. A large block may be used to illustrate this idea. Work is not done on the object if the object does not move. Moreover work is not done on an object that is held in midair despite the fact that in order to hold the object, a force must be applied. Work *is* done on the object by an applied force only if the object moves in the direction of that applied force. Note that work is done only when the object accelerates (i.e., $F \propto a$). No work is done on an object moving at a constant velocity. Since work is a vector quantity, it can be negative. "Negative" refers to the direction work is being done. For example, a person may lift and then lower a block. In this case, the person does positive work, while gravity does negative work of the same magnitude.

In certain cases it can be difficult to specify the source of the force that causes the work. For example, it is often said that no work is done when an object is carried across the room. If the object accelerates, then work is done! If the object is held, grabbed or gripped, then the person applies a force in the direction of the movement. This results in work being done by that person. If the object rests on a surface, such as a hand, table or tray, these surfaces exert an upward force. This force is perpendicular to the movement of the object and does not cause motion. Therefore it is not the cause of work being done. In this case, it is friction that provides the source of the force. Friction is unidirectional with the acceleration of the object.

Work can also be related to energy. It takes work to produce a change in energy and vise versa.

Materials

Wood block or a box, large, with "M" written or painted on the side

Safety Precautions

Although this demonstration does not pose any obvious risks, always follow laboratory safety rules.

Demonstration

Direct the students' attention to the wooden block or box. Discuss the concept of a mass and what it would take to move the object. Next, apply a force on the block by lifting it up. Ask the class if you have done any work so far. Ask students what the term "work" means.

After discussing the physical definition of work, hold the block in midair. Ask students if you are doing any work on the block. Remind them that work is done only when the object accelerates.

Next, lower the block and ask students if work is done on the block, and if so, who or what did the work. In this case, gravity does negative work on the block.

Raise the block to the level of your chest and carry it across the room with it resting on the palm of your hands (do not grip the block). Again, ask students if *you* are doing any work on the block. Emphasize that work is done on the object by an applied force only if the object moves in the direction of the applied force.

Friction Is Working

In many cases friction is the source of the force that does work.

Application | Work • Friction • Newton's First Law • Newton's Third Law

Theory

Work is done on an object by an applied force only if the object moves in the direction of the applied force. In certain cases it can be difficult to specify the source of the force that causes the work. In this demonstration, friction provides this force. If the object rests on a surface, such as a hand, table or tray, then the surface exerts an upward force. This force is perpendicular to the movement of the object and does not cause motion. Therefore it is not the cause of work being done. Friction is unidirectional with the acceleration of the object.

There are many examples of friction as the basis of work. The frictional force of automobile brakes acts in the opposite direction of the vehicle, causing it to decelerate—constituting negative work. Sliding into home base, skidding to stop, and the air resistance of a parachute are other examples of negative work being done by friction. Friction can also provide positive work as in the case of walking. As you take a step, you apply a backward-directed force during which friction provides an equal and opposite force that propels you forward. The turning wheels of an automobile, pushing off skates or skis, and a forward jump are also examples of positive work being done by friction. These last examples may also be used as an illustration of Newton's Third Law.

Materials

Audio-visual (AV) cart, the type used for an overhead projector

Dynamics cart, low friction

Dynamics track, low friction

Preparation

Place a track on an AV cart and then place a dynamics cart on the track. Be sure the track is level. If necessary, use paper or cardboard to shim the track.

Demonstration

Ask students: What will happen if the AV cart is moved across the room? Those students who have a good understanding of Newton's First Law will say that the dynamics cart will stay still while the track moves beneath it. Verify students' responses by moving the AV cart a meter across the room. The friction between the wheels and the track is not sufficient to change the position of the dynamics cart. Ask students why the dynamics cart stays still. Discuss the role that friction plays in the motion of objects. Ask students if any work was done on the dynamics cart. Discuss the requirements of physical work.

Turn the cart on the track over so that the wheels are facing upward and repeat the demonstration. The dynamics cart should move with the track since the friction is sufficient to change the position of the cart. Ask students if any work was done on the cart. Discuss the requirements of physical work in terms of an applied force causing an object to accelerate through a distance. Emphasize that it is friction that moved the dynamics cart across the room.

A bow is used to illustrate that work is a linear function of force versus distance.

Application | Work • Energy graphing • Hooke's Law • Elasticity • Springs

Theory | An archer's Recurve bow behaves like a spring. Drawing the bow requires a steadily increasing force. A graph of the force versus the draw of the bow indicates a linear relationship between the two. This relationship is known as Hooke's Law : $F = kx$, where k represents the spring constant. Drawing the bow results in the storage of spring potential energy, $U_e = \frac{1}{2}kx^2$. In order to store the energy, work must be done on the bow by applying a force, F, to cause the draw, e.g., $W = Fx$. The area under the force versus distance graph is equal to the work done on the bow.

All linear springs exhibit the same characteristics as the bow in this demonstration. Physical therapists use elastic resistance bands to assist patients in rehabilitating muscles, ligaments and joints. As in any spring, these bands require a greater force as they are increasingly stretched. Bungee cords also exhibit this property.

Not all elastic materials and springs are linear-elastic or "Hookean" materials. Modern car shock absorbers are not linear-elastic by design since they dissipate energy during their operation. Other materials are Hookean under certain conditions or within limiting ranges. Rubber is generally considered a non-Hookean material since it is stress- and temperature-dependent.

Materials | Bathroom scale

Dowel, hardwood, $24'' \times \frac{5}{8}''$

Handsaw, small

Plywood, $8'' \times 8'' \times \frac{3}{4}''$

Recurve bow, with a 20+ lb draw (a compound bow will not work)

Sandpaper, 60–80 grit, 1 sheet

Tape, electrical (vinyl tape), $\frac{1}{2}''$, bright-colored

Wood glue

Safety Precautions | An archer's bow should never be drawn and released quickly without an arrow. Otherwise, the stored energy will be absorbed by the bow and may cause it to shatter. Bows should never be drawn with arrows in a classroom or other confined area. The bow in this demonstration should be relaxed slowly once drawn out to prevent injury or damage to the equipment.

Preparation

Drill a ⅝″ diameter hole one-half inch into the center of a 8″ × 8″ × ¾″ piece of plywood. Glue a 24″ long, ⅝″ hardwood dowel into this hole. Glue a piece of 60–80 grit sandpaper to the bottom of the apparatus. Use a handsaw to cut a ⅛″ deep notch into the center of the top end of the dowel to serve as a channel for the bow string. Use vinyl tape to mark a series of 5-cm increments along the length of the dowel. The zero-mark should be located at the point where the relaxed bow rests against the dowel as illustrated in the diagram.

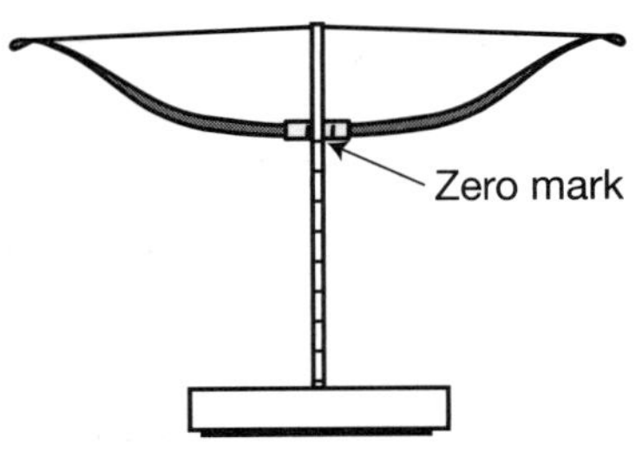

Demonstration

Ask students to describe the force needed to draw a bow. Does the force remain constant or does it increase? Place the assembly squarely onto a bathroom scale. Have a student balance the bow on the dowel as illustrated in the diagram. Note the force indicated on the bathroom scale at this point; subtract this amount from subsequent readings. Instruct the student to pull the bow down to the first 5-cm mark and then ask her to call out the reading on the bathroom scale. Write this force versus distance data in a table on the board. Have the student continue to pull on the bow, calling out the force readings at each subsequent marking.

Plot the relative units of force (lbs) per distance (cm). Direct students' attention to the fact that the relationship between force and distance is linear. Ask students if any work was done on the bow. After discussing the various ideas, ask them to propose a method of determining how much work was done. Finally, show students how to determine the work done on an object using a force versus distance graph.

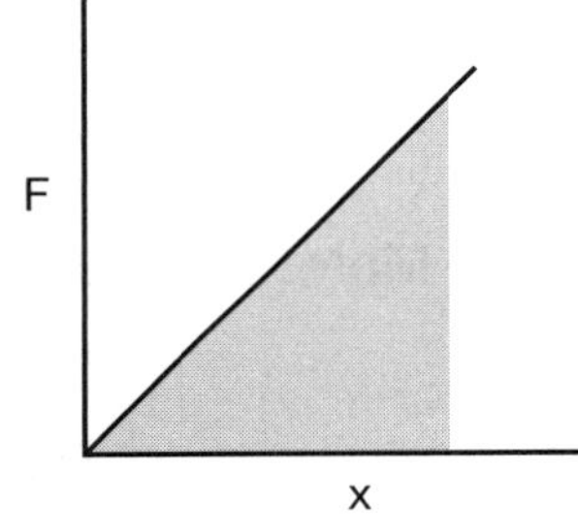

Reference

Robinson, Paul. "Wrap Your Energy in a Bow;" *Conceptual Physics Laboratory Manual;* Prentice Hall: Needham, Massachusetts, 2002; p 85.

The potential energy due to the mass of a pile driver is used to drive a nail into a piece of wood.

Application	Work • Potential energy • Kinetic energy • Work–energy theorem • Friction
Theory	A pile driver is a device used to drive piles into soils to serve as the foundation of a building structure. Piles are typically made from timber, steel, reinforced concrete, and pretensioned concrete columns. The pile driver consists of a large mass which is raised by a pulley system and then allowed to fall freely along a pair of guiding rails onto the pile. The mass is made of a cylinder of iron known as the driver. As the mass is lifted, it gains potential energy which in turn is converted into kinetic energy that is then transferred to the pile, causing it to be driven into the soil further and further with each successive drop. The friction between the surfaces of the pile and the soil is what the force of the driver is designed to overcome. A pile driver may be effectively used to present the work–energy theorem since the potential energy, PE, of the driver provides the work that drives the pile against the friction force, F_f, over a distance, d:

$$PE = KE = W = F_f \cdot d$$

Materials	Hammer

Mass, 500–1000 g, with a diameter $< 1\frac{1}{2}''$

Permanent marker, fine-tip

Roofing nail, 2–3″

Ruler, cm, small

Transparent fluorescent light disposal tube, 1 m
(available from home improvement stores)

Wood block, small, soft (example: pine, $2'' \times 4'' \times 4''$)

Safety Precautions	Wear safety glasses when performing this demonstration. Although this demonstration does not pose any obvious risks, always follow laboratory safety rules.
Preparation	Use a fine-tip permanent marker to make marks every 1 cm along the length of a roofing nail, beginning from the pointed tip.
Demonstration	Use a hammer to drive a 2–3″ roofing nail into a piece of soft wood to a depth of 1 cm; that is, up to the first mark on the nail. Ask students what resists the nail as it is driven into the wood. Then ask students how can the friction between the nail and the wood be determined. After discussing various ideas, explain how a pile driver works.

Place a 1-m long transparent fluorescent light disposal tube over the nail embedded in the wood. Drop a 500–1000 g mass into the tube onto the nail. Remove the mass from the tube and continue dropping the mass onto the nail until the nail has been driven into the wood an additional 1 cm. Calculate the work done on the nail and then calculate the friction between the nail and the wood.

An analysis of the graph of the friction force versus the distance the nail moves emphasizes how work is the integral of force times distance.

Reference

Sutton, Richard Manliffe. *Pile Driver in Demonstration Experiments in Physics.* McGraw–Hill: New York, 1938; M-133, p 60.

Pop Goes the Energy

A bowl-shaped piece of rubber is dropped and bounces higher than the release point.

Application	Potential energy • Conservation of energy Spring constant • Coefficient of restitution
Theory	Dropping an object provides the object with an initial gravitational potential energy. The greater the height, the greater the gravitational potential energy stored. Mass also affects the potential energy of an object. When an object is dropped, it will rebound no more than the original release height. The height that the object actually reaches is dependent upon the coefficient of restitution, COR. The COR measures the change in velocity as a fractional value of the object before and after the collision with the surface. A COR value of 1 indicates that the collision is perfectly elastic, whereas a COR value of 0 indicates that the collision is inelastic, that is, the objects stick together upon collision. The original Super Ball® first introduced in 1965 had a COR of 0.89–0.92, meaning the ball would return to 92% of its initial height. A material that is stressed by tension or compression will have a natural tendency to return to its relaxed point. The tension or compression serves as a store of energy called spring potential energy. In this demonstration, the dropped object rises above its drop position due to the spring potential energy given to the bowl-shaped piece of rubber upon folding it. The bowl-shaped piece of rubber begins with both gravitational potential energy and spring potential energy. When released, the gravitational potential energy is converted to kinetic energy. Upon contact with the surface, the spring energy is released. Both the spring potential and the initial kinetic energies are then converted into the final kinetic energy, causing the object to travel upwards. Although the COR of the material may be less than 1, the object surpasses the initial drop height because of the spring potential energy.
Materials	Racquetball Sharp scissors Utility knife *Optional:* A commercial toy often called a "Rubber Popper Toy" can be used in this demonstration.
Safety Precautions	Use caution whenever working with a utility knife. Wear safety glasses when performing this demonstration. Always follow laboratory safety rules when performing demonstrations.

<table>
<tr><td>Preparation</td><td>Use a utility knife to carefully bisect a racquetball along its seam. Two poppers can be made from a single racquetball. Invert one of the halves by pushing the convex side inwards and then return it to its original state. Notice how difficult it is to bring the ball to its relaxed state. Use a pair of sharp scissors to trim away the edge of the ball (approximately 3–5 mm) to such a degree that the inverted ball will pop from its flexed configuration to its relaxed configuration when dropped from a height of 30–40 cm. Cutting away too much of the material will cause it to pop immediately. When drop-testing, hold the flexed device horizontally with the inverted end pointed upward.</td></tr>
<tr><td>Demonstration</td><td>Drop the racquetball bowl from a height of about 30–40 cm without inverting it. The bowl will rebound, as expected, to a height of about half of the release point. Discuss the exchange of energy that occurs in the system. Point out that some of the energy is lost and converted to heat and sound. Then invert the racquetball bowl and drop it from a height of about 30–40 cm. The ball should bounce up to a height of 1 m or more. Discuss the exchange of energy that occurs in this system. The increase in height is due to the energy that was stored in the inverted ball prior to release.</td></tr>
<tr><td>Reference</td><td>Becker, Robert. "Racket Ball"; Twenty Demonstrations to Knock Your Socks Off; Flinn Scientific: Batavia, IL; Vol I, p 72.</td></tr>
</table>

A spring is used to launch balls of different masses illustrating the work done by a spring.

Application	Work • Spring potential energy • Spring constant • Hooke's Law
Theory	A spring is a device made of an elastic material that can store energy called spring potential energy, PE_s. Springs can be compressed, stretched, bent (compressed and stretched) or twisted (bent around an axis). In all these cases, energy is stored due to the temporary deformation of bonds between the atoms in the material. The stiffness of a spring is directly related to the amount of energy a spring can store and is measured by the spring constant, k:

$$PE_s = \tfrac{1}{2} k x^2$$

where x is the amount the spring is stretched or compressed from its equilibrium point. The change in x can also be described by Hooke's Law:

$$F = kx$$

where F is the force that stretches or compresses the spring.

In this demonstration, when the compressed spring is released, the spring potential energy is transferred into the ball, giving it kinetic energy that is then converted to gravitational potential energy:

$$\tfrac{1}{2} k x^2 = \tfrac{1}{2} m v^2 = mgh$$

Note that the greater the mass of the ball, the smaller the launch velocity of the ball, and the lower the height achieved.

Materials

Adhesive, construction or plumber's epoxy

Compression spring, $\frac{3}{4}'' \times 2''$; $k = 300\text{–}500$ (available at hardware stores)

Golf ball

Ping-Pong ball

Plywood, $\frac{3}{4}'' \times 5'' \times 7''$

Safety Precautions

Wear safety glasses when performing this demonstration. Always follow laboratory safety rules.

Preparation

Use construction adhesive to secure the spring vertically in the center of a piece of plywood.

Demonstration | Place a golf ball onto the spring apparatus, press down fully, and then release. Depending on the spring constant, the ball should launch upward about 30–50 cm. (It may take some practice to have the ball launch straight up).

Ask students what will happen if a less massive ball is used in this demonstration. The Ping-Pong ball is expected to launch upwards 2 m. Consider discussing how Hooke's Law, gravitational potential energy, and spring potential energy relate in this situation.

A bowling ball pendulum is released from the nose of a volunteer and permitted to swing back and forth.

Application | Conservation of energy • Potential energy • Kinetic energy • Pendulum

Theory | The pendulum gains gravitational potential energy when raised above its rest point, which is the equilibrium point and the lowest point in its swing. Upon release, the potential energy is converted into kinetic energy. At the bottom of the swing, all of the potential energy has been converted into kinetic energy. Hence, the maximum potential energy (and absence of kinetic energy) of the system exists at the upper position of the pendulum while the maximum kinetic energy (and absence of potential energy) exists at the bottom of the swing. This demonstration illustrates the law of conservation of energy in that the pendulum will never rise above its release point. In fact, the pendulum never even reaches the release point because some of the initial potential energy is lost to friction in the mounting system of the pendulum, air resistance, and the microscopic degradation of the cable as it is stretched.

Materials

Bamboo skewer

Bolt snaps with ½″ swivel eye, 2

Bowling ball, any size

Chair, rigid-back, non-rolling

Concrete block

Drill

Drill bit, 3/16″

Epoxy, construction grade

Eye bolt permanently mounted to ceiling

Screw eye, ¼″ × 3″

Wire rope (aircraft cable), 3/16″ diameter, general purpose steel

Wire rope clips, 3/16″, 4

Wire rope thimbles, 3/16″, 2

Note: A permanently ceiling-mounted eyebolt is required for the demonstration. The bowling ball apparatus should never be hung from a suspended ceiling. The ceiling hook must be anchored to a rigid structure such as a concrete ceiling or steel beam.

Safety Precautions | The pendulum apparatus should never be left unattended and should not be operated by anyone other than the instructor. It is imperative that the ceiling hook is securely anchored into a rigid structure such as a concrete ceiling or steel beam. The wire rope should be inspected before each use of the demonstration apparatus to ensure that the wire clips are secure. The path of the pendulum during the use of the demonstration must be free of any obstacles. In this demonstration always release the bowling ball without giving it a push.

Preparation

Drill a ¼″-inch hole three inches into the bowling ball. Use a bamboo skewer or similar item to coat the inside of the hole with construction grade epoxy. Coat the threads of the screw eye in a similar fashion. Use a large pliers or screwdriver for leverage in inserting the screw eye fully into the bowling ball. Allow epoxy to cure per manufacturer's directions.

The length of the cable should be such that it is long enough to form a 45° angle between the nose of the seated volunteer, the ceiling hook, and the rest position of the pendulum. When considering the placement of the ceiling hook, be sure that the path of the pendulum will be free of any obstacles.

Pass the open end of the thimble through the ring of the swivel eye of the bolt snap. Thread cable through two clips and around the thimble and back through the clips. Secure the clips to the wire using the appropriate size wrench or socket. Repeat on opposite end of cable.

Clip one side of the cable to the ceiling eyebolt, and suspend the bowling ball from the other end of the cable. Lift the pendulum to the release position and let go to verify the safety of the pendulum before attempting the demonstration.

Demonstration

Ask the student to sit in a chair facing the suspended bowling ball with his or her back firmly pressed against the back of the chair. The chair must be positioned so that the bowling ball will reach the individual's nose. Make sure that the path of the swing of the pendulum is free of any obstacles. Bring the bowling ball up to the individual's hands and ask the student to just touch his nose with the bowling ball. Ask the class to predict what will happen if the individual releases the ball from this position. The volunteer must also be instructed not to move after the ball  is released. After a short discussion, ask the volunteer to release the ball—*without pushing!* To ensure safety, the instructor should watch for and prevent the person from making any forward movement after the release. Have the students explain the motion of the released bowling ball in terms of energy. Emphasize that a system cannot have any more energy than was initially put into it. In this demonstration the energy of the system depends on the initial height of the ball, as this will relate to its total potential energy. The magnitude of the kinetic energy of this system can be illustrated by placing an object, such as a concrete block, at the bottom of the swing of the pendulum.

Disposal

Place broken block in trash.

Reference

Head, James H. "Faith in Physics: Building New Confidence with a Classic Pendulum Demonstration"; *The Physics Teacher;* 1995; vol 33, pp 10–15.

A ball is released down one end of a parabolic ramp and returns to the same height on the opposite end of the ramp.

Application

Conservation of energy • Potential energy • Kinetic energy

Theory

In all kinematic processes, energy is always conserved. In this demonstration a ball is positioned above the bottom of the track providing the ball with gravitational potential energy. Upon release, the potential energy is converted to kinetic energy. At the bottom of the track, the kinetic energy is at a maximum while the potential energy is equal to zero. The kinetic energy of the ball causes it to move up the other side of the track until the kinetic energy is completely converted back into gravitational potential energy. In an ideal system, the final height reached by the ball will be equal to the release height of the ball:

$$mgh_i = \tfrac{1}{2} mv^2_{(bottom)} = mgh_f$$
$$h_i = h_f$$

Although the total energy in this system is conserved, the ball does not actually reach the same height as the release height because some of the initial energy is consumed by the work done by friction and the ball's rotation. But, the design of this apparatus is such that the difference between the two heights will be difficult to measure. The difference may be more pronounced if a lighter ball, such as one made of wood or plastic, is substituted for the metal ball.

A roller coaster must be lifted to a particular height in order to provide it with enough potential energy to propel the coaster through the rise and fall of the ride. Since the maximum potential energy is given to the coaster at the first rise, successive hills cannot be higher than the first.

Materials

Coffee can, large

Hacksaw

Rubber ball, 2″, with a low coefficient of restitution (e.g., a "sad ball")

Screw gun

Steel bookshelf tracking, 8′, available at home hardware stores

Wood glue

Wooden board, 1″ × 3″ × 4′

Wooden boards, 1″ × 3″ × 24″, 2

Wood screws, 2″, 6

Wood screws, ¾″, 6

Wear safety glasses when performing this demonstration. Always follow laboratory safety rules.

Preparation

Attach the shorter boards to the ends of the longer board using wood glue and screws. Pre-drilling the holes will prevent the wood from splitting. Using a large coffee can as a form, bend the shelving track about 140° one-third from one end of the track. Be careful not to bend too far to prevent the track from kinking. Use the short wooden screws to secure the track to the wooden frame, bending and/or trimming the ends of the track as necessary.

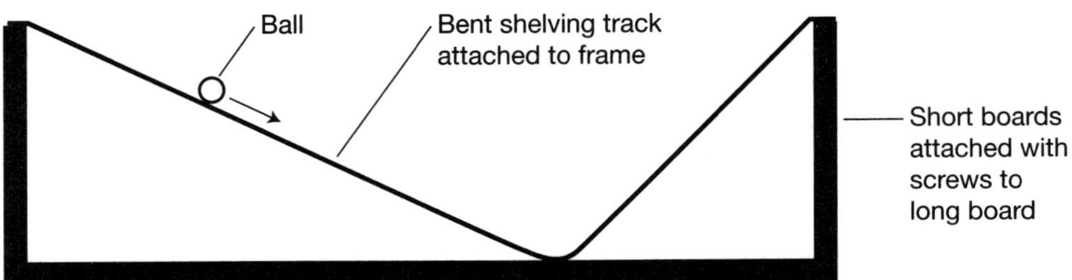

Demonstration

Position the ball halfway up the longer side of the ramp. Ask students to predict what will happen if the ball is released. After a short discussion, release the ball to verify the students' hypothesis. Repeat the demonstration with the ball released from different heights.

Finally, ask students if there is a position on the track from which can be launched the ball so that it will fly off the opposite side of the track. After discussing students' responses, release the ball to test their hypotheses. At the conclusion of the demonstration, emphasize that the ball cannot travel higher than its release height.

Two or more students run up a staircase to determine who is more powerful.

Application | Power • Work • Gravitational potential energy • Force

Theory | Power is the rate of doing work or consuming energy. In this demonstration, the maximum power outputs of a number of volunteers are compared by having them running up a fixed height. The work done is equal to the product of the weight and distance the body is lifted:

$$P = \frac{W}{t} = \frac{mgh}{t}$$

Most likely the heavier student will need to exert the most power because of his or her relative mass. However, you may have a lighter student that can move fast enough to overcome the mass difference, which would result in a lighter person exerting more power than the heavier individual.

Modern aerobic exercise machines have built-in calculators that determine the power output of the user based on the mass inputted by the individual. The power output of a hydroelectric dam is dependent upon the change of the gravitational potential energy of the water per unit of time.

Materials | Bathroom scale

Meter stick

Staircase or exercise stepper

Stopwatch

Safety Precautions | Be certain that the staircase is not slippery and that the volunteers wear well-fitting athletic footwear.

Demonstration | Invite two or more students to assist with the demonstration. Volunteers must wear appropriate footwear. Begin by measuring the weight of each volunteer using a bathroom scale. Convert the weights measured to mass units (kilograms). Walk to a nearby staircase. Ask one student to measure the height of one step and then calculate the height of the entire staircase. Have the first volunteer stand at the base of the staircase. Instruct this student to run up the staircase as fast as possible, without skipping steps. Use a stopwatch to measure the time taken for the ascent. Repeat the process using other volunteers. Return to the classroom and have students calculate the power exerted by each individual. Usually, the heavier individual will exert more power.

Note: An exercise stepper may be used in place of a staircase. In this case have the volunteer step up and down 10 times. Keep in mind that the student does work only in the upward direction, while gravity does negative work in returning the student back to the floor. Since the student is working only half of the total time, divide the total time of the activity in half when calculating the power.

Stop That Pendulum!

Interrupting a pendulum at the bottom of its swing illustrates the law of conservation of energy.

Application	Pendulum • Potential energy • Kinetic energy
Theory	The amount of potential energy in a pendulum is directly related to the height at which it is released above the bottom of the pendulum's swing, i.e., *mgh*, with the mass of the pendulum and gravity being constants. When the pendulum reaches the bottom of its swing after being released from height *h,* the energy is all kinetic energy: ½ mv^2, yet is equal to the original potential energy of *mgh.* If, at this point, you are to interrupt the swing of the pendulum by placing a rod in the way of the supporting string of the pendulum, the pendulum must still return to the original release height as the mass and gravity have not changed and energy must be conserved.
Materials	Adhesive, construction or epoxy Board, magnetic, chalk or white Neodymium magnets, ½″ × ½″ disc, 2 Nylon string, thin, about 3′ (e.g., fly-fishing line) Pendulum bob, metal, 1″ (or a golf ball with a hole drilled through it) Wooden dowels, ½″, 6″ length, 2
Safety Precautions	Keep magnets away from magnetic media and all electronic equipment. Always follow all laboratory safety rules when performing demonstrations.
Preparation	Use epoxy or construction adhesive to attach the magnets to the ends of the dowels. After allowing the glue to set, connect a pendulum bob to the middle of one dowel using a 3′ piece of thin nylon string. Pendulum length can be adjusted to the size of the magnetic board. Place the pendulum support rod towards the top of board, making sure it can swing freely and not come in contact with anything during its swing. With the pendulum at rest, draw a vertical line on the board behind the pendulum string and down to the bob. Now, decide how high you'd like to lift the pendulum in the demo, and draw a horizontal line from this point across the board to the other side where the pendulum will swing.
Demonstration	Lift the pendulum up to the horizontal line and allow it to swing freely. Direct students' attention to the fact that the pendulum rises approximately to the same level at which it was released, indicating that energy is conserved. Discuss the conversion from potential energy to kinetic and back again as the pendulum swings freely. With the pendulum at rest, place the second rod a third of the way from the bottom of the pendulum swing on the vertical line behind the string. Ask students what will happen in this case if the pendulum is raised to the horizontal line and

released. After discussing the possible scenarios release the pendulum to show that the bob returns to the same height from which it was released on the other side of the swing, regardless of the interruption. Discuss the conversion of energy, its different forms, and why conservation of energy predicts what is observed.

Reference Sutton, Richard Manliffe. "Pendulum Board," *Demonstration Experiments in Physics;* McGraw-Hill: New York, 1938; M-132, p 60.

Slow the Block Down

A block of wood is launched as a result of the swing of a pendulum onto a surface upon which friction slows the block to rest.

Application

Gravitational potential energy • Kinetic energy • Negative work • Friction

Theory

Friction is a force that opposes motion. When friction is opposing the motion of an object, it does negative work. In this demonstration a block that is resting on a pendulum is launched along the surface of a table and eventually comes to rest due to the work done by friction. The situation may be analyzed in terms of energy and work in that the gravitational potential energy given to the block by raising it on the pendulum is consumed by the work done by friction:

$$mgh = F_f \cdot d = \mu mgd$$

Consequently, the coefficient of friction is merely dependent upon the initial height of the pendulum and the distance the block slides across the table:

$$\mu = h/d$$

Materials

Cup hooks, 2	Short rod, $\approx$ 30 cm long
Drill	Table clamps, large, 2
Drill bits, ½″ and ⅛″	Table rods, 1 m long, 2
Dowels, ½″ × 3′, 2	Wooden board (preferably hardwood), 2″ × 3″ × 5″
Plywood, ¾″ × 5″ × 5″	Wood glue
Right angle clamps, 2	

Safety Precautions

Wear safety glasses when performing this demonstration. Always follow laboratory safety rules.

Preparation

Drill a ⅛″ pilot hole into one end of both dowels. Place a drop of wood glue into each hole and screw in a cup hook. Then drill a ½″ hole ½″ deep into the plywood board as illustrated in the diagram.

Place a drop of glue into these holes and insert the two dowels. Orient the cup hooks in the same direction so the device may be suspended from a horizontal bar.

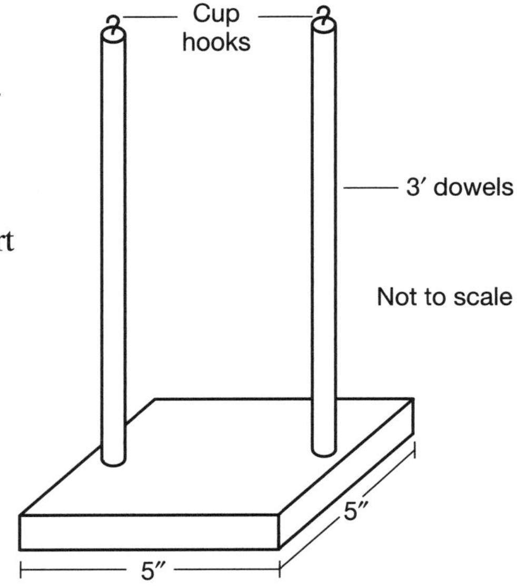

"""

Secure two table clamps with vertical rods to the edge of a demonstration table. Use two right angle clamps to attach a 30-cm horizontal rod to vertical poles. Suspend the apparatus from the horizontal rod so the upper surface of the plywood is even with the table surface as illustrated in the diagram.

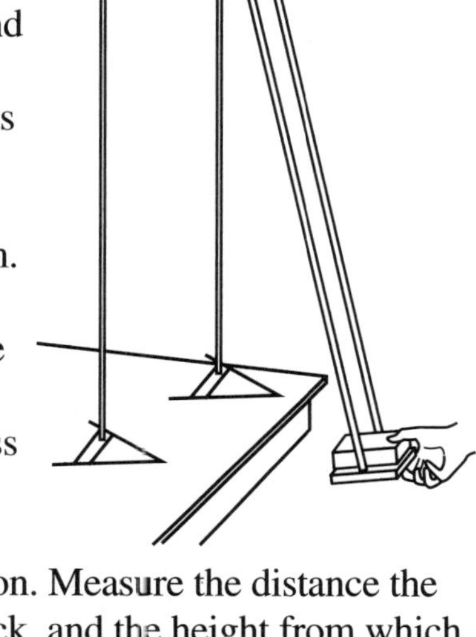

Demonstration

Place a hardwood block onto the base of the pendulum. Pull the pendulum back about 30° from the rest point and then release it. Upon collision with the edge of the table, the block should slide across the table. Ask students why the block did not continue to move across the table. After discussing how friction brought the block to rest, ask students what else needs to be measured in order to calculate the work done by friction. Measure the distance the block traveled using a meter stick, the mass of the block, and the height from which the block was released. Then have students calculate the coefficient of friction between the block and the table. Emphasize that friction is doing negative work in slowing the block down.

Notes

Chapter 6

Momentum

Egg in Sheet

An egg is thrown into a bedsheet as an illustration of impulse.

Application Impulse • Change in momentum • Collisions

Theory It takes an impulse (J) to change the momentum (Δp) of an object. The impulse acting on an object refers to the product of the force acting on the object and the time the force acts.

$$J = F\Delta t = \Delta p = m\Delta v$$

In this demonstration an egg is thrown into a hanging bedsheet. Whether the egg is thrown against a wall or the sheet, the same force is needed to bring the egg to rest. If the force is applied to the egg quickly (low Δt), as is the case when thrown against a wall, the impulse is large enough to break the egg. If the duration of the force is increased (large Δt), the impulse is lower and the egg may survive the collision. Automobile air bags use this same principle in that the air bag is designed to distribute the impact force over time. During a bungee jump, the bungee cord stretches over time and therefore decreases the impulse exerted on the jumper as she is brought to rest. If the cord were simply a rope, then the effect of the impulse would not be dissimilar to simply hitting the ground. Protective sports equipment, such as padding, gloves and helmets, also distribute the force over time as they are compressed and thus reduce the impulse acting on the individual.

Materials Bedsheet, full-size

Egg, raw

Safety Precautions Although this demonstration does not pose any obvious risks, always follow laboratory safety rules.

Demonstration Select an open area for this demonstration, away from hanging obstructions or other obstacles. Invite two students to assist with the demonstration. Have them hold a bedsheet so it hangs like a screen, but with the bottom of the sheet pulled upwards forming a pocket. The top of the sheet should be pulled tight, but the bottom should be yielding. There should be at least a meter of space between the back of the sheet and the wall or any other object. Invite a third volunteer—preferably one that has some experience pitching a ball—to throw an egg into the hanging bed sheet. Position this student about 3–5 meters away from the sheet and instruct him or her to aim for the center of the sheet. Instruct the student to throw the egg with enough force so the egg hits the sheet. Repeat the experiment a few more times—each time increasing the force of the throw. No matter how hard the egg is thrown, it should not break when it hits the sheet. Discuss how the bedsheet is used to decrease the impulse of the collision.

The impulses of the collision of two objects with different coefficients of restitution are compared.

Application	Impulse • Coefficient of restitution • Inelastic collisions
Theory	You may be familiar with the commercially available "Happy and Unhappy Balls." Sold in pairs, these balls are relatively identical in appearance but are made of two different materials. The happy ball is made of neoprene while the sad ball is made of polynorbornene (Composition may vary among manufacturers.) When dropped from the same height, the happy ball bounces as expected, while the unhappy ball barely rebounds. Slicing the sad ball open reveals that the core of the ball is composed of an aerated rubber-like material, which contributes to the ball's lack of bounce. The height that the balls actually reach is dependent upon the coefficient of restitution, COR. The COR measures the change in velocity as a fractional value of the object before and after the collision with a surface. A COR value of 1 indicates that the collision is elastic, whereas a COR value of 0 indicates that the collision is inelastic. Only the collisions between atoms or molecules exhibit a COR = 1. In the case of the happy ball, the high COR results in a higher return velocity than that of the unhappy ball. The total force acting on the block in this demonstration, which causes it to topple, is derived from two sources: the force due to the forward movement of the ball, plus the push-off force due to the rebound of the ball. Since the happy ball has a greater rebound, it exerts a greater force on the block, causing it to topple. The COR of protective sports equipment, such as padding, gloves and helmets, is low and therefore minimizes the impulse on the individual. The COR value of racquetballs, baseballs and basketballs are regulated in order to standardize the equipment used in the games. Imagine being struck by a racquetball or baseball. Which would you rather be struck with—one with a high COR value or a low one?

Materials

Fly fishing line, ≈ 1 m

Happy/Unhappy Balls, 2 pairs

Hot glue

Hot glue gun

Right-angle clamp

Ring stand, tall

Ruler, 30-cm

Screw eyes, ¼″, 2

Wooden blocks, $1″ \times 3″ \times 5″$, with square and smooth cut ends, 2

Wooden dowel, ½″, 12″ long

Safety Precautions

Although this demonstration does not pose any obvious risks, always follow laboratory safety rules.

Preparation

Attach ¼″ screw eyes to each ball. Attach 50 cm of fly fishing line to each ball to make a pendulum. Tie one pendulum to the end of a wooden dowel and the second spaced about 15 cm away from the first. Be sure the lengths of the pendulum are equal. Use hot glue to secure the fishing line to the dowel. Use a fine-tipped permanent marker to draw an "x" marking the center of gravity of one face of each block. Use a right angle clamp to secure the dowel apparatus to a ring stand. Position the blocks so the balls are just touching the mark on the center of each block.

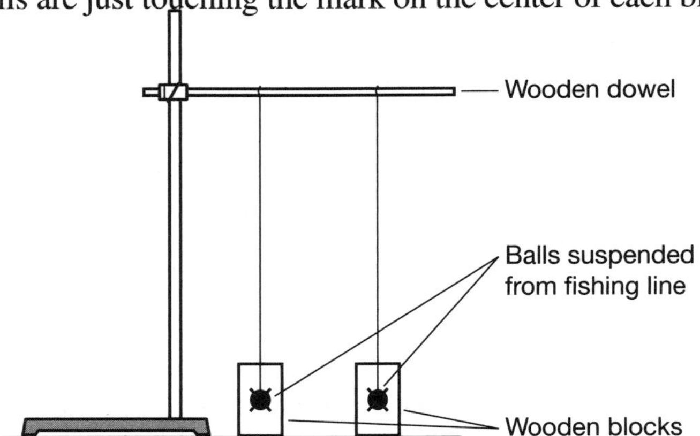

Demonstration

Drop a pair of Happy/Unhappy Balls onto a table from a height of about 1 meter. The happy ball should bounce as expected, while the unhappy ball will barely rebound. Ask students if they can propose a reason for the difference. Perhaps discuss the property of the material's coefficient of restitution (COR). Next, without identifying the ball type, raise the happy ball pendulum to the specific height needed to just knock the block over. Use a ruler to measure the height. Raise the unhappy ball to the *same* height and release. The block should not be toppled by the unhappy ball. Ask students to identify which of the pendulums is happy or unhappy. Discuss the relationship between the COR value, impulse, and the force imparted to the block.

A small ball is stacked onto a larger one and dropped onto a surface resulting in the smaller ball bouncing with a greater velocity.

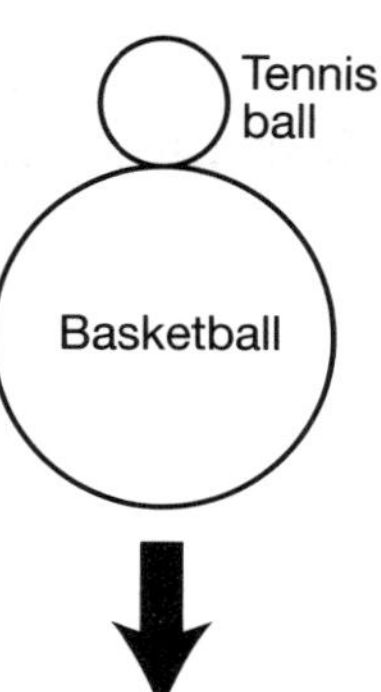

Applications | Conservation of momentum • Inelastic collisions • Explosion

Theory | In this two-part demonstration, two balls having different masses are dropped together onto the floor, resulting in a transfer of momentum. In the first case, the balls are in contact with one another during the fall. In simple terms, upon collision with the floor, the initial momentum, p, of the system is split between the two masses. Since the tennis ball has a much lower mass, m, than the basketball, the final velocity, v, of the tennis ball will be greater, causing it to rise much higher than the basketball:

$$\Sigma p = (m_T + m_B)v_i = m_T v_{T_f} + m_B v_{B_f}$$

The separation of the system is called an explosion. In terms of momentum, an explosion is the reverse of a *completely* inelastic collision.

The tennis ball separates from the basketball due to the compression wave created when the basketball collides with the floor. This compression wave travels from the point of collision through the basketball, causing the basketball to dimple outward towards the tennis ball. This ejects the tennis ball from the contact point between the two balls, as if the basketball struck the tennis ball. For this reason, the balls separate.

The demonstration is repeated, but this time instead of dropping the balls while they are in contact with one another, the system is dropped so there is a small space (2–3 cm) between the balls. In this case, the demonstration is an example of an inelastic collision, since both balls fall independently of one another. The inelastic collision occurs when the upward moving basketball collides with the downward moving tennis ball. The final velocity of the tennis ball is tremendous compared to the first demonstration and is due to two factors: the momentum transferred from the larger basketball to the tennis ball and the fact that the tennis ball changes direction.

$$\Sigma p = m_B v_{B_i} - m_T v_{T_i} = m_T v_{T_f} - m_B v_{B_f}$$

Materials | Basketball

Tennis ball

Hard floor surface or tabletop (carpeted floor is not desirable)

Note: To achieve ideal results, a new tennis ball should be used and the basketball should be fully inflated.

<table>
<tr><td align="right">Safety
Precautions</td><td>Do not conduct this demonstration under light fixtures or other obstacles. Although this demonstration does not pose any obvious risks, always follow laboratory safety rules. Please be careful when dropping the balls; do not throw them.</td></tr>
<tr><td align="right">Demonstration</td><td>Drop the basketball from chest height onto the floor or tabletop to see how high it bounces. Repeat with the tennis ball for comparison. Ask students to predict what would happen if the tennis ball were placed on top of the basketball and the balls were dropped together while in contact with one another. After a short discussion, drop the system onto the floor. Upon collision with the floor, the tennis ball should rebound to at least a meter above your height, while the basketball bounces about half the height as when it was bounced alone. Discuss the transfer of momentum from the system to the two balls upon explosion. Repeat the demonstration, but this time instead of dropping the balls while they are in contact with one another, drop the system so there is a small space (2–3 cm) between the balls. The tennis ball should be ejected with a much higher velocity and strike the ceiling. However, the basketball will achieve a much lower height than it did previously when stacked. In fact, the upward-moving basketball will change direction upon colliding with the tennis ball at a height slightly higher than the diameter of the basketball. Discuss why the balls behave differently in this case.</td></tr>
<tr><td align="right">Reference</td><td>Spradley, Joseph L. "Velocity Amplification in Vertical Collisions"; American Journal of Physics; American Association of Physics Teachers; 1987; vol 55, pp 183–184.</td></tr>
</table>

A ball is launched into a series of balls arranged on a curved track causing the last ball to be ejected as a model of an elastic collision.

Application   Model of elastic collision • Transfer of momentum
Conservation of energy • Kinetic theory

Theory   This demonstration is a close approximation of what occurs during an elastic collision—but only an approximation. It may be used to illustrate the concept of an elastic collision, however, it must be emphasized that this collision is actually inelastic with a minimal amount of kinetic energy lost. According to the kinetic theory, *only* the collisions between atoms or molecules are elastic. In an elastic collision, both momentum and kinetic energy, *KE*, are conserved. Macroscopic collisions result in the deformation of the objects (even if slight), which reduces *KE*. The sound and heat generated during collision also consume *KE*. It is worth emphasizing that macroscopic collisions are either inelastic or completely inelastic—they are not elastic! When the collision is completely inelastic, the objects stick together. An explosion is the reverse of a completely "inelastic collision." Be aware of the fact that some physicists use the term "perfectly elastic" (i.e., "it really is elastic") when they really mean elastic and use the term "elastic collisions" when speaking of examples that *approximate* elastic collisions as described above. The term perfectly elastic should be avoided to prevent confusion. There are three types of collisions: elastic, inelastic, and completely inelastic (or *perfectly* inelastic).

In this demonstration a single ball is given gravitational potential energy by raising it up one side of a curved ramp. The ball is then released, causing it to travel down the ramp and collide with a line of four similar balls—all initially at rest. The initial collisions result in a series of collisions from one ball to the next, resulting in the final ball being launched up the opposite side of the ramp. This ball will rise to nearly the same height as the release point of the original ball, thus illustrating the conservation of energy. In terms of momentum, the series of collisions may be viewed as a series of momentum transfers between adjacent balls. If a released ball is replaced with a ball identical in size but made from a denser material, such as lead, then the resulting momentum transfer will launch the last ball with a greater velocity. The greater velocity will result in this ball being launched off the track. The greater velocity is due to the difference in masses between the first and final ball.

A question arises—why do two balls eject two balls? Assuming that the balls have the same mass, can two balls eject one ball, but with a greater velocity? Momentum is conserved in all collisions. If two balls did eject one ball, then

$$mv_i + mv_i = mv_f, \text{ or } 2v_i = v_f$$

That is, the final velocity of the single ball would have to be twice the velocity of the two initial balls. In terms of kinetic energy,

$$\frac{1}{2}mv_i^2 + \frac{1}{2}mv_i^2 = \frac{1}{2}mv_f^2$$

By substituting the momentum equation into the kinetic energy equation, we find that the final kinetic energy would exceed the initial kinetic energy of the system, which is impossible:

$$\frac{1}{2}mv_i^2 + \frac{1}{2}mv_i^2 \neq \frac{1}{2}m(2v_i)^2$$

$$\text{or,} \quad v_i^2 + v_i^2 \neq (2v_i)^2$$

In all cases, any number of balls will eject an equal number of balls.

Note: There are three types of collisions: elastic, in which both momentum and kinetic energy are conserved; inelastic, in which momentum is conserved but kinetic energy is not; and, finally, completely (or perfectly) inelastic, in which the objects stick together and momentum is conserved but kinetic energy is not. (Perfect in this sense refers to a complete collision or unification.) Some authors have invented their own terms to describe collisions, which creates confusion among students. For example, some authors subdivide elastic collisions into two groups: elastic and perfectly elastic, classifying billiard ball collisions as elastic and those between atoms and molecules as perfectly elastic. The idea is well-meaning, but from the position of an educator, the choice of words ("perfectly") is unfortunate since students will confuse the two types of "perfect" collisions. In the case of "perfectly inelastic" collisions, the objects sticks together; perfect = sticks together. In the case of a "perfectly elastic" collision, the collision is "really" elastic not simply elastic; perfect = it actually is. (Whatever that means!) Furthermore, the term elastic has been used to describe collisions between atoms or molecules for more than two centuries ever since the kinetic theory was first proposed. There is no such thing as "perfectly" elastic. The collision is either elastic or it isn't! It is best to adhere to mainstream terminology described above. That is, there are three types of collisions: elastic, inelastic and completely inelastic (or *perfectly* inelastic).

<table>
<tr><td>Materials</td><td>Band saw, jigsaw or hand coping saw</td></tr>
</table>

Materials

Band saw, jigsaw or hand coping saw

Bookshelf track, 24″ (available at home hardware centers)

Lead ball, 1″

Ring shank drywall nails, 1″, 8

Smooth wooden boards, 1″ × 3″ × 24″, 2

Steel balls, 1″

Wood screws, 1″, 5

Note: A commercially available apparatus commonly referred to as "Newton's Cradle" may be used in this demonstration.

Review manufacturer's recommendations before working with power tools. Wear safety glasses when operating power tools. Although this demonstration does not pose any obvious risks, always follow laboratory safety rules. Use caution when launching the balls, as they may fall off or become airborne.

Preparation

Sketch a shallow curve along one edge of a $1'' \times 3'' \times 24''$ smooth wooden board. The depth of the curve should be about $1''$. Cut out this curve with a band saw, jig saw or coping saw. Mount a piece of shelving track to the curved edge using ring shank drywall nails. Use wooden screws to mount the curved piece to a second wooden board that will serve as the base of the apparatus.

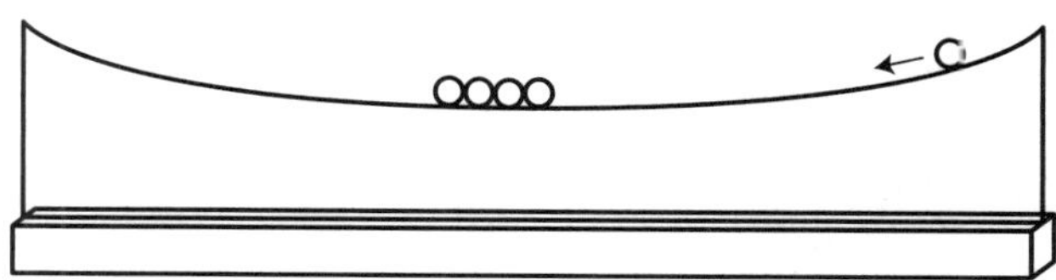

Demonstration

Direct students' attention to the ball collision device and ask them what will happen if one ball is raised to the top of one side of the track and released. After discussing various ideas, raise and release the ball. It is expected that the ball farthest from the point of collision will be launched up the opposite side of the ramp—nearly to the top. Discuss the conservation of energy in this situation. Next, ask students what will happen if two balls were released. After discussing various ideas, raise and release the two balls. It is expected that two balls on the opposite side will be launched up the ramp. As in the case of a single ball, the two balls will reach nearly the same height as the pair of released balls. Discuss why the two balls do not launch only one ball, rather than two up the opposite side of the ramp. Ask students what will happen if a ball, with the same diameter yet made from a denser material, were launched into the other balls. After discussing various ideas, raise and release the ball. Finally, try launching 3 or 4 balls to see what happens. Point out that if the ball does not transfer its energy to another, it will keep oscillating back and forth along the track. To prevent misconceptions, emphasize that the collisions approximate the elastic collisions that occur between atoms or molecules, but the collisions observed in this demonstration are really inelastic.

Rocket Balls

A simple magnetic accelerator apparatus is used to propel a steel ball as a twist on the Classic Newton's Cradle (or Newton's Trough) demonstration.

Application

Transfer of momentum • Conservation of momentum
Inelastic collisions • Magnetic force

Theory

Newton's Cradle or Trough is often used to illustrate the transfer of momentum during a collision. The source of energy for the system is the initial energy of the first ball upon its release. Using this device, the expected outcome is that the velocity of the outgoing ball is nearly equal to the velocity of the incoming ball, illustrating the conservation of momentum. In this demonstration a magnet is placed in front of the target balls supplying an additional force which will accelerate the incoming ball. The final momentum of the ball depends not only on the velocity of the incoming ball, but it also depends upon the impulse, $F\Delta t$, provided by the magnet:

$$p_2 = F\Delta t + p_{1_i}$$

$$p_2 = ma\Delta t + p_{1_i}$$

$$p_2 = m\frac{\Delta v}{\Delta t}\Delta t + p_{1_i}$$

$$mv_2 = m(\Delta v) + mv_{1_i}$$

$$v_2 = (\Delta v) + v_{1_i}$$

where Δv is the increase in velocity of the incoming ball due to the magnetic force.

Magnetic fields used in a particle accelerator boost the velocity of subatomic particles. In some cases a series of magnetic accelerators are used in amusement park rides to provide an initial velocity to a roller coaster car.

Materials

Adhesive, construction

Bookshelf track, 36″ (available at home hardware centers)

Neodymium disk magnets, 1″ D × ½″ L, with poles on face, 3

Smooth wooden board, 1″ × 2″ × 36″

Soft drink can, empty

Steel balls, 1″, 7

Safety Precautions

Neodymium magnets of any size are very powerful. Exercise caution when separating magnets or bringing magnets near one another as they can potentially pinch skin or collide together and emit broken shards. Keep magnets away from magnetic media including computer discs, audio/video tapes and cards with magnetic strips (e.g., credit cards, swipe cards, etc.)

<table>
<tr><td align="right">Preparation</td><td>Using construction adhesive, mount a 36″ piece of shelving track along the center of a 1″ × 2″ × 36″ wooden board.</td></tr>
</table>

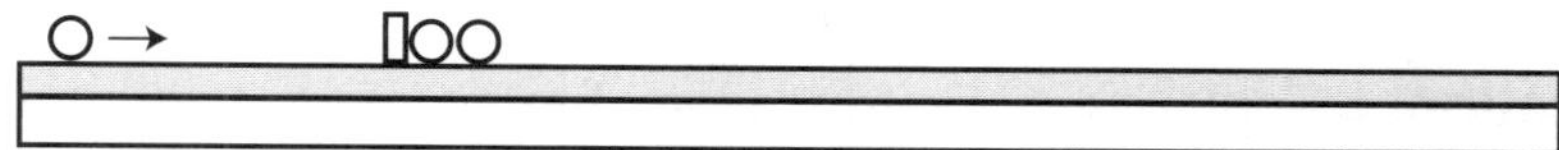

<table>
<tr><td align="right">Demonstration</td><td>Place two 1″ steel balls and one neodymium magnet about 15 cm away from one end of the track, as illustrated in the diagram. Orient the magnet with its poles aligned horizontally. Ask students what will happen if a third ball is rolled into the magnet–ball system. After discussing various ideas, slowly roll the ball towards the magnet. The ball is expected to crash loudly into the magnet, causing the ball on the opposite end of the system to be launched with a speed much faster than the initial speed of the first ball. Most likely students would not have expected this result, so repeat the demonstration and ask students to observe the movement of the balls very carefully. Direct students' attention to the fact that the momentum of the launched ball must be greater than the initial momentum of the rolled ball. Discuss how this could be possible. Perhaps remind students that it takes an impulse to change the momentum of a system. Discuss how the magnet provides an additional force that contributes to the change in momentum and consequently increases the speed of the launched ball.</td></tr>
</table>

To increase the effect observed in this demonstration, try adding two more sets of balls and magnets, each separated about 10 cm apart.

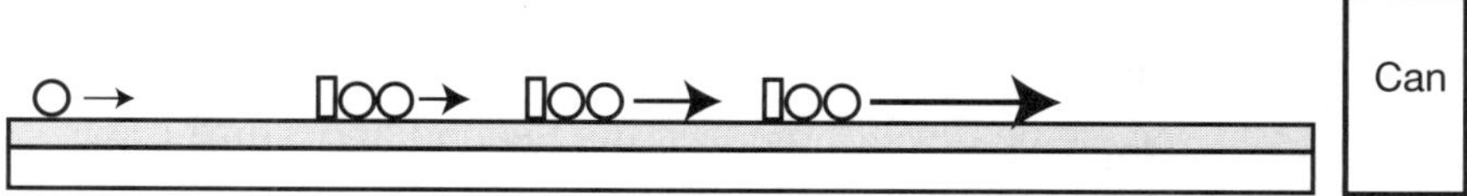

Also place an empty soft drink can at the end of the track. Colliding with the can will result in a loud noise that will emphasize the force of the collision.

<table>
<tr><td align="right">Reference</td><td>Special thanks to Dave Barnes of Arbor Scientific for sharing this idea.</td></tr>
</table>

Exploding Carts

A pair of stationary dynamics carts is forced apart by a spring in order to observe the conservation of momentum.

Application

Explosion • Conservation of momentum • Newton's Third Law

Theory

An object at rest has zero momentum, p. If such an object were to explode into two or more pieces, the total momentum of all pieces would still be equal to zero. Although the momentum of each piece is equal, the velocity of each piece is dependent upon its mass. The greater the mass, the smaller the velocity.

$$\text{Before explosion:} \quad \Sigma p = 0 = (m_1 + m_2)v = (m_1 + m_2)(0)$$

$$\text{After explosion:} \quad \Sigma p = 0 = m_1 v_1 + m_2 v_2$$

where vectors v_1 and v_2 must be in opposite directions.

In the case where $m_1 = m_2$, $v_1 = -v_2$. In terms of momentum, an explosion is the reverse of a completely inelastic collision.

A rocket lifting off is an example of a controlled explosion during which the relatively smaller mass of propellant quickly exits the engine while the more massive rocket rises up slowly. Bombs and other explosions throw out pieces of shrapnel whose velocity is dependent upon their individual masses. In the explosive demolition of a building, demolition experts must take into account the potential movement of individual components of the building when placing their charges.

Conservation of momentum can also be viewed as an example of Newton's Third Law. In fact, it was Newton's understanding of the conservation of momentum that led to the development of his Second and Third Laws.

Materials

Balance or spring scale

Dynamics cart masses (each equivalent to the mass of 1 cart), 2

Dynamics carts, 2 (one with spring)

Small hammer or similar-weight object

Track with magnetic end bumpers

Note: An air track system may be used to obtain similar results.

Safety Precautions

Although this demonstration does not pose any obvious risks, always follow laboratory safety rules.

Demonstration | Use a balance or spring scale to show students that the carts and masses are equivalent in mass. Cock the spring in the spring-loaded cart and position it facing the other cart on the center of the dynamics track. The carts should be in contact with one another. Ask students what will happen to each cart if the spring is released. After discussing the various ideas, use a small hammer or similar device to release the spring. The carts should move apart with equal velocities and arrive at the ends of the tracks at the same time, repelling off the magnetic bumpers and returning to their initial position. Next, double the mass of one cart by adding a cart mass to it.

Ask students to predict what will happen if this system is exploded apart. After discussing the various ideas, release the spring. The lighter of the two carts is expected to travel twice as fast as the heavier cart, showing that the velocities of the carts are dependent upon the relative masses of the carts. Emphasize that this illustrates the conservation of momentum. Repeat the demonstration using a second mass to triple the mass of one cart.

Fire When Ready

A corked plastic bottle filled with alcohol vapor is exploded, resulting in the bottle and cork moving in opposite directions.

Application

Explosion • Conservation of momentum • Newton's Third Law

Theory

A rocket lifting off is an example of a controlled explosion during which the relatively smaller mass of propellant quickly exits the engine while the more massive rocket rises up slowly. This illustrates the principle of conservation of momentum. In this demonstration an alcohol–air mixture contained in a corked plastic bottle is ignited, creating an explosive expansion of gases that propels the cork in one direction while the bottle is pushed in the opposite direction. This demonstration can also be explained in terms of Newton's Third Law.

Materials

Alcohol, 5 mL (≥90% ethanol or methanol (not rubbing alcohol, which is 70% isopropyl alcohol); available from your school's chemistry department

Cork, #4 (because of its greater mass, do not use a rubber stopper)

Cup hooks or ½″ screw eyes, 2

Duct tape

Graduated cylinder, 10-mL

Hot glue, low temperature

Hot glue gun, low temperature

Nylon string (mason's string), ³⁄₁₆″ diameter

Paper clips, large, 2

Pin, large

Soft drink bottle, plastic, 2-L

Straw, plastic

Tesla coil or Piezo BBQ Igniter with a pair of alligator patch cords

Safety Precautions

Alcohol is highly flammable and its vapors are explosive; keep away from open flame. This demonstration intentionally uses an alcohol vapor–air mixture as a fuel source. Refrain from using more than the specified amount. Only the vapor is necessary. Excess unvaporized alcohol will burn, potentially creating an additional hazard. Keep hands and face away from the path of the bottle rocket and moving cork. Always launch the bottle–cork system horizontally in front of your audience and parallel to them so that neither bottle nor cork is directed towards the audience or yourself. Wear safety goggles or a face shield when performing this demonstration. Please review current Material Safety Data Sheets for additional safety, handling, and disposal information.

Use a pin to pierce a hole on either side of a 2-L plastic soft drink bottle, about 5 cm from the base. Unfold two large paper clips and insert one clip into each hole, pushing them in far enough so that the wire ends are about 1 cm apart from one another inside the bottle. Use *low* temperature hot glue to secure the wire to the bottle and to keep the wires in line with one another. These wires will serve as a sort of spark plug that will allow the ignition of the fuel mixture. Thread a piece of mason's string through a plastic straw. If using a flexible straw, trim off the flexible component prior to threading the string through it. Mount the string tightly to cup hooks attached to opposite walls in the room such that the string is parallel to your audience. Use duct tape to secure the modified bottle to the straw.

The system must be designed so neither the bottle nor the cork will move toward the audience. The vicinity of the cup hooks should be free of breakable items such as glass

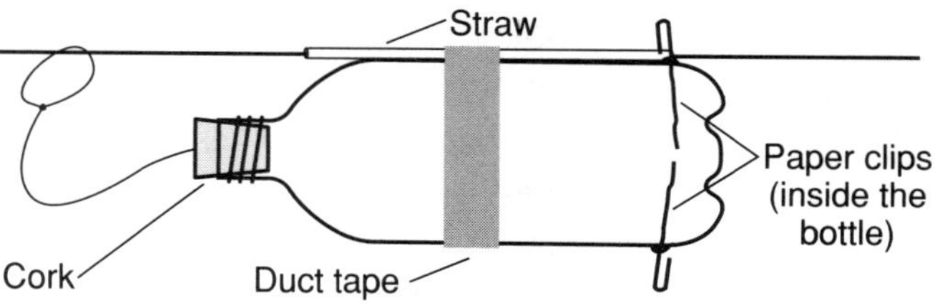

cabinet doors. Use hot glue to secure a 20-cm long piece of mason's string to the end of a #4 cork. Tie the other end of the string around the horizontally mounted guideline creating a large loop so the cork and string can slide freely along the guideline.

Add 5 mL of alcohol to the plastic bottle, swirl around a few times to vaporize the liquid and then pour the excess alcohol down a sink drain. Flush the drain with water. Use duct tape to secure the bottle to the straw mounted on the guideline as illustrated in the diagram. Holding the bottle with one hand, insert the cork securely, but not too tight, only using your thumb to push on the cork. Slide the bottle–cork apparatus to the middle of the guideline.

Ask students what will happen if the fuel mixture is ignited inside the bottle using a spark. After discussing the various ideas, use a Tesla coil or piezo igniter (see below) to ignite the mixture. The exploding vapors will create a sudden expansion of gas that will increase the pressure inside the bottle until the cork is ejected in one direction while the bottle moves in the opposite direction. Take the opportunity to discuss Newton's Third Law and emphasize that this is not a cause and effect event, but rather a force pairing—the movement of the cork and bottle occurs simultaneously. Direct attention to the fact that upon ignition, the low mass cork has a higher velocity than the high mass bottle.

Piezo Igniter: An inexpensive igniter system for this demonstration may be fashioned from a commercial BBQ igniter. (Piezo pushbutton BBQ igniters are available at hardware stores.) Simply solder a pair of patch cords with alligator clips to the leads of the igniter. Wrap the igniter with electrical tape and slip the wires and igniter through a 6″ length of ¾″ PVC pipe.

Fuel Me Up

A water rocket is launched with and without water to explore the effect of mass on an object's momentum.

Application	Explosion • Conservation of momentum • Newton's Third Law
Theory	A rocket lifting off is an example of a controlled explosion during which the relatively smaller mass of propellant quickly exits the engine while the more massive rocket rises up slowly. This illustrates the principle of conservation of momentum. In the case of the toy rocket, when the mass of the propellant is increased, but the velocity of the propellant remains the same, the velocity of the rocket will increase proportionally. In this demonstration, the mass of the propellant is changed by using either air only or a water–air combination. If the pressure in the rocket is the same in both cases, then the water–air fueled rocket will rise to a higher elevation due to its greater mass. This demonstration can also be explained in terms of Newton's Third Law.
Materials	Commercial pump-action water rocket (available at toy stores) Water
Safety Precautions	Keep hands and face away from the path of a rocket. Always launch the rocket away from your audience. This demonstration must be performed outside. Do not try it in the classroom or inside any building.
Preparation	Prior to class, fill the water rocket half-full with water and determine the maximum number of pumps that it will take for an average student to pressurize the rocket.
Demonstration	This demonstration must be performed in an open area outside, such as an athletic field. Discuss how the water rocket operates. Explain that the rocket is normally filled with water. Ask students what will happen if the rocket is launched with or without the water, but pressurized with air in both cases. During the discussion mention that in the first case, air will be used as the propellent, while in the second case, air *and* water will serve as the propellant. Invite a student to assist with the demonstration. Have her pump the rocket full of air with the predetermined number of pumps measured prior to class. In order for the experiment to be controlled, the number of pumps must be the same with and without water. Have the student launch the rocket upward, away from herself and the audience. Have the class estimate the maximum apogee of the rocket. Next, have the student fill the rocket with water and again pressurize it with the same number of pumps previously used. Launch the rocket, again taking care to direct it away from the student and all onlookers, and have students compare this apogee to the previous launch. Note that the water-filled rocket will have a much greater velocity and consequently poses an increased hazard.
	The phenomenon may be further explored by varying the amount of water in the rocket but maintaining the same number of pumps.

Two students on Rollerblades push off one another as an illustration of conservation of momentum.

Application | Explosion • Conservation of momentum • Newton's Third Law

Theory

In this demonstration two students on Rollerblades push off one another, resulting in the two students moving away from one another. In terms of momentum, p, the process is considered an explosion:

$$\Sigma p = 0 = m_A v_{A_f} - m_B v_{B_f}$$

Although the initial velocity of both students is zero, momentum is conserved due to the opposite direction of their velocities upon explosion. The relative velocities of the students depend only upon their relative masses.

The demonstration may also be viewed in terms of Newton's Third Law in that the action of one student is matched by the reaction of the other student, regardless whether one or both students push off one another. If one student pushes, they both move apart; if both students push, they also move apart. The speed with which they move apart depends upon the total force exerted as well as the masses of the students.

Materials

Hallway or large floor area, free of obstruction

Rollerblades, well-fitting, pair, 3

Helmets, well-fitting, 3

Safety padding set, 3

Note: The rollerblades must have nearly the same coefficient of friction. Two students should be similar in mass, while the third should be lighter or heavier by 50 percent.

Safety Precautions

Select an area such as a hallway or a large room free of obstructions, such as desks, chairs, glass cabinets or lab tables. Be certain that the entire floor area is dry and free of dirt and debris. Students must be appropriately outfitted with safety helmets and padding to protect them in the event of a fall. Do not allow students to rollerblade except for the motion needed for the demonstration. Students without prior experience on Rollerblades should not participate as volunteers in this demonstration.

Demonstration

Invite three students with rollerblading experience to participate in the demonstration. Two students should be similar in mass, while the third should be lighter or heavier by 50 percent. Have each student don a helmet, appropriate padding, and a pair of Rollerblades. Instruct students to remain seated until they are needed to perform the demonstration. Invite the two volunteers with similar mass to stand facing each other in front of the audience, with their backs to the sides of the demonstration area. Ask students what will happen if the two volunteers push off one another. After discussing the various ideas have the two students hold their palms out to each other and push with equal force. Instruct students to stand with their backs and hips locked and knees bent to prevent their bodies from being rotated backwards. As an added safety feature, ask two additional students to move alongside and slightly behind the rollerblading students to catch the students in case they lose their balance. The students are expected to move away from one another at the same speed, demonstrating that momentum is conserved.

Ask the students to repeat the demonstration, but this time instruct one student to provide the push, while the other simply stays still. Discuss the fact that momentum is still conserved. The observation can also be viewed as a demonstration of Newton's Third Law—action and reaction forces.

Next, pair up a heavier student with a lighter one. Ask students how the pair will move when they push off one another. Following a short discussion, confirm their predictions. The lighter student is expected to move faster than the heavier student, once again illustrating the conservation of momentum. This result occurs regardless of who pushes, whether one or both students.

Reference Etkina, Eugenia; Maiullo, David. *Physics on Rollerblades* (video); Rutgers University: New Brunswick, NJ, 1997.

A student wearing Rollerblades pushes on a wall, illustrating both conservation of momentum and Newton's Third Law.

Application

Conservation of momentum • Newton's Third Law • Explosion

Theory

In accordance with Newton's Third Law, when a person pushes on a wall, the wall pushes back. The understanding is incomplete without considering the conservation of momentum of the system. In terms of momentum, p, when a person pushes on a wall and moves away from the wall, the wall actually moves away from the person.

$$\Sigma p = 0 = m_\text{P} v_{\text{P}_f} - m_\text{W} v_{\text{W}_f}$$

The velocity of the wall is far less than the velocity of the person due to the difference in mass, and the wall's movement is far less obvious. In this demonstration, having the person on Rollerblades provides an opportunity to observe the velocity of the person pushing on the wall. The movement of the wall is made observable by employing a laser–mirror system (see Lazy Wall, p 6 for a complete explanation). The laser–mirror system takes advantage of the change in geometry of the reflected laser beam, similar to how a seismometer amplifies ground movement.

Materials

Angle clamp, adjustable

Hot glue

Hot glue gun

Laser pointer

Mirror, glass, small, flat (e.g., 2 cm × 2 cm)

O-ring to fit around laser pointer

Ring stand

Student outfitted fitted with Rollerblades, helmet and safety padding

Safety Precautions

Select an area free of obstructions such as desks, chairs, glass cabinets or lab tables. Be certain that the floor is dry and free of dirt and debris. The student volunteer must be appropriately outfitted with a safety helmet and padding to protect against injury in the event of a fall. Do not allow students to rollerblade except for the motion needed for the demonstration. Students without prior experience on Rollerblades should not participate as volunteers in this demonstration. Never shine laser beams into eyes.

Preparation

Select a wall with a surrounding area that is free of obstructions such as desks, chairs, glass cabinets or lab tables. Be certain that the entire floor area is dry and free of dirt and debris. Mount a small mirror at chest level on

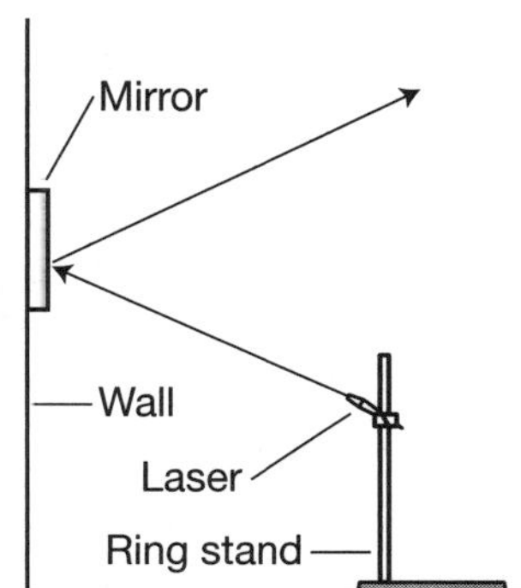

this wall using hot glue. Slip an O-ring over a laser pointer to keep the laser turned on. (Some laser pointers have a constant on switch, which makes this step unnecessary.) Use an adjustable angle clamp to mount the laser to a ring stand. Position the ring stand on the floor about a meter away from the wall. Adjust the laser so it reflects off the mirror onto the center of the ceiling, making the dot visible to all.

Demonstration

Turn on the laser and adjust its beam so it is reflected off the middle of the mounted mirror onto the ceiling. Explain to students that the beam will serve as an indicator of wall movement—similar to how a seismometer works. Invite a student with rollerblading experience to participate in the demonstration. Have the student don a well-fitting helmet, appropriate padding, and a pair of Rollerblades. Instruct the student to remain seated until he or she is needed to perform the demonstration. Ask students what will happen if the volunteer on Rollerblades pushes on the wall. After discussing the various ideas, have the student push on the wall near the mounted mirror. As the volunteer pushes on the wall, the reflected laser beam will move as both the wall and the volunteer move away from one another. Discuss the fact that the movement of the reflected beam demonstrates that the wall actually does move. Point out that since the wall is much more massive than the student, its movement is minimal in comparison.

Push Me, Throw You

A student throws a medicine ball while standing on Rollerblades, illustrating conservation of momentum.

Application	Explosion • Conservation of momentum
Theory	In this demonstration a student on Rollerblades throws a medicine ball away from herself, resulting in the student and the ball moving away from one another. In terms of momentum, p, the process is considered an explosion:

$$\Sigma p = 0 = m_A v_{A_f} - m_B v_{B_f}$$

Although the initial velocity, v, of both objects is zero, momentum is conserved due to the opposite direction of their velocities. The relative velocities of the objects depend only upon their relative masses.

Materials

Hallway or large floor area, free of obstructions

Medicine ball (available from the athletic department)

One student outfitted with Rollerblades, helmet, and safety padding

Safety Precautions

Select an area such as a hallway or large room free of obstructions including desks, chairs, glass cabinets or lab tables. Be certain that the entire floor area is dry and free of dirt and debris. The student must be appropriately outfitted with a well-fitting safety helmet and padding to protect her in the event of a fall. Do not allow students to rollerblade except for the motion needed for the demonstration. Students without prior experience on Rollerblades should not participate as volunteers in this demonstration.

Demonstration

Invite one student with rollerblading experience to participate in the demonstration. Have her don a well-fitting helmet, appropriate padding, and a pair of Rollerblades. Have the volunteer remain seated until she is needed to perform the demonstration. Invite the volunteer to stand in front of the class facing to the side of the demonstration area. As an added safety feature, ask an additional student to move alongside and slightly behind the rollerblading student to catch her in case she loses her balance. Hand a medicine ball to the volunteer. Ask students what will happen if the volunteer throws the medicine ball away from her body. After a short discussion, have the student throw the ball away from the center of her body toward a cleared area. The student is expected to move in the opposite direction than the medicine ball was thrown. Have students compare the relative velocities of the ball and the volunteer.

What's My Speed?

The principle of a completely inelastic collision is used to determine the pitching speed of a softball.

Application | Completely inelastic collision • Conservation of momentum
Friction • Conservation of energy

Theory | The speed of a softball may be determined indirectly by measuring the work done by friction and by using the conservation of kinetic energy and the conservation of momentum. In this demonstration the momentum of a softball is transferred to a box–ball system during this completely inelastic collision. A completely inelastic collision is one during which the colliding objects stick together:

$$\Sigma p = m_{ball} v_{ball} = (m_{ball} + m_{box}) v_i \quad \text{or} \quad v_{ball} = \frac{(m_{ball} + m_{box}) v_i}{m_{ball}}$$

where v_i represents the velocity of the box–ball system.

During a completely inelastic collision, momentum is conserved but kinetic energy is not. In this demonstration most of the kinetic energy lost is consumed by the work done by friction causing the box–ball system to come to rest:

$$W_{friction} = \Delta KE_{ball+box}$$

An insignificant amount of kinetic energy is lost to the deformation of the paper and the sound and heat produced during collision.

In order to achieve the final result (v_{ball}), the coefficient of kinetic friction must first be determined:

$$F_f = \mu m_T g \quad \text{or} \quad \mu = \frac{F_f}{m_T g}, \text{ where } m_T = (m_{ball} + m_{box})$$

Once the coefficient of friction is known, then the velocity of the box–ball system may be determined by equating the work done by friction to the initial kinetic energy of the box–ball system:

$$(\mu m_T g) \cdot x = \tfrac{1}{2} m_T (v_f^2 - v_i^2) \quad \text{or} \quad v_i = \sqrt{-2 \mu g x}$$

where x refers to the distance the box slides after collision.

Finally, the law of conservation of momentum may be used to calculate the initial velocity of the ball:

$$(mv)_{ball} = m_T \sqrt{-2 \mu g x} \quad \text{or} \quad v_{ball} = \frac{m_T \sqrt{-2 \mu g x}}{m_{ball}}$$

<table>
<tr><td align="right">Materials</td><td>

Balance
Baseball
Box, cardboard, large (e.g., $24'' \times 24'' \times 24''$)
Masking tape
Meter stick
Newspaper
Spring scale, 20 N
Softball
Utility knife
Optional: Softball with built in speedometer

</td></tr>
<tr><td align="right">Safety Precautions</td><td>

Although this demonstration does not pose any obvious risks, always follow laboratory safety rules. The demonstration area should be clear of breakable items such as glass, electronic devices, and other fragile materials. Keep the audience behind the pitcher when she throws the ball.

</td></tr>
<tr><td align="right">Preparation</td><td>

Use a utility knife to remove the flaps from the opening of a large box. Crumple single sheets of newspaper and fill the box.

</td></tr>
<tr><td align="right">Demonstration</td><td>

Review the principles outlined in the theory section of this demonstration with students as a preparation for the activity. Explain how the speed of a softball may be determined indirectly by measuring the work done by friction and using conservation of kinetic energy and the conservation of momentum.

Move your audience into a hallway or to a large floor area that may be used to conduct this demonstration. Begin by measuring the mass of the box and softball separately using a spring scale and/or balance. Place the box on its side and use a spring scale to determine the force of kinetic friction by dragging the box containing the softball across the floor with a constant speed. The spring scale must be held horizontal to ensure the correct value. Use this information to calculate the coefficient of kinetic friction.

Ask a student who has experience in pitching a softball or baseball to assist with the demonstration. Place the box on its side with the open-end facing the student. Use masking tape to mark the location of the box on the floor. While standing at a distance about 5 m away from the box, have this student throw the softball into the newspaper-filled box. The box should slide down the hallway a few meters. Use a meter stick to measure the distance the box slides. The ball must only contact the newspaper and the box must slide straight, otherwise some energy will be lost. Add more crumpled newspaper and repeat if necessary. Return to the classroom and calculate the initial speed of the ball.

For further exploration consider comparing the experimental result obtained here with the result obtained using a softball with a built-in speedometer.

</td></tr>
</table>

Curl It This Way

Two commercial air pucks are used to illustrate the principles behind 2-D collisions.

Application | Glancing collisions • Model of elastic collision
Conservation of momentum in 2-D

Theory | Collisions often occur in two and three dimensions. Games such as billiards and curling, and high-energy experiments in atomic and nuclear physics, during which subatomic particles collide, are all examples of two- and three-dimensional collisions. Atomic and nuclear collisions are classified as elastic collisions since energy is conserved. While collisions between two billiard balls or between two curling stones can serve as reasonable models of elastic collisions, these examples are actually inelastic since energy is lost through deformation, sound, and heat. The actual amount of energy lost is low so it is reasonable to ignore.

In this demonstration a toy air puck collides with a stationary air puck, resulting in both pucks moving away from the point of collision at separate angles. The conservation of momentum can be demonstrated.

Since the velocities of both pucks are known before and after the collision, this problem is relatively simple.

According to the law of conservation of momentum:

$$p_{1y_{(i)}} = p_{1y_{(f)}} + p_{2y_{(f)}}, \text{ assuming } p_{2y_{(i)}} = 0$$
$$m_1 v_{1y_{(i)}} = m_1 v_{1y_{(f)}} \cos\theta_1 + m_2 v_{2y_{(f)}} \cos\theta_2$$

where θ is the angle measured with respect to the horizontal.

In this demonstration, $m_1 = m_2$, so the equation is reduced to:

$$v_{1y_{(i)}} = v_{1y_{(f)}} \cos\theta_1 + v_{2y_{(f)}} \cos\theta_2$$

Furthermore, the momentum is also conserved in the *x*-direction:

$$0 = p_{1x_{(f)}} + p_{2x_{(f)}} \quad \text{where} \quad p_{1x_{(i)}} = 0 \quad \text{and} \quad p_{2x_{(i)}} = 0$$
$$0 = m_1 v_{1x_{(f)}} \sin\theta_1 + m_2 v_{2x_{(f)}} \sin\theta_2$$

Since $m_1 = m_2$,

$$0 = v_{1x_{(f)}} \sin\theta_1 + v_{2x_{(f)}} \sin\theta_2 \quad \text{or} \quad v_{1x_{(f)}} \sin\theta_1 + v_{2x_{(f)}} \sin\theta_2$$

Materials | Beanbags, 2

Kick Dis™ toys, 2

Large open floor space such as gymnasium floor, $\approx 10 \text{ m} \times 10 \text{ m}$

Measuring tape

Protractor

Stopwatches, 2

Tape, masking

Although this demonstration does not pose any obvious risks, always follow laboratory safety rules.

Preparation

Charge your Kick Dis puck battery packs overnight. Prior to class, test the Kick Dis to verify that the toys slide smoothly over the floor surface. Use masking tape to create a 30 cm × 30 cm x–y coordinate axis in the center of the demonstration area. Mark a spot exactly 5 m below the axis with a small piece of masking tape.

Demonstration

This demonstration must be performed on an open floor space such as a cafeteria or gymnasium, at least 10 m × 10 m in size. Select five volunteers to assist with the demonstration. Provide two volunteers with stopwatches, two volunteers with bean-bags, and the last volunteer with a Kick Dis puck.

Place one Kick Dis puck at the axis of the grid on the floor. Position two students at the 5 m mark below the axis—one student with a puck and the other with a stop-watch. Have a third student with a stopwatch stand 5 m above the axis. Instruct the fourth and fifth volunteer to stand in the first and second quadrant holding a bean-bag in their hands.

Outline the roles of each student and their responsibilities in data collection:

1. Student 1, beginning slightly below the 5-m mark below the axis, is to push the puck towards the axis and then release it just prior to the 5-m mark to provide the puck a constant velocity as it travels within the grid.

2. Student 2 is to use the stopwatch to determine the time it takes for puck 1 to travel from the 5-m mark to puck 2, which is located at the axis.

3. Student 3 is to start timing at the start of the collision, and then yell out "Time!" once three seconds is reached. (This time interval depends upon the push on the puck provided by student 1—adjust as needed).

4. Students 4 and 5 are to drop a bean bag at the spot reached by respective pucks when they hear student 3 call-out "Time"!

Ideally, the trajectory angles upon collision should be between 30° and 60°. If not, repeat the demonstration.

Use a tape measure to measure the distance traveled by each puck after collision. Then use a protractor and the tape measure to measure the angle of trajectory of each puck with respect to the y-axis.

Return to the classroom and calculate the momentum of the system before and after collision to verify that momentum is conserved in a 2-D collision.

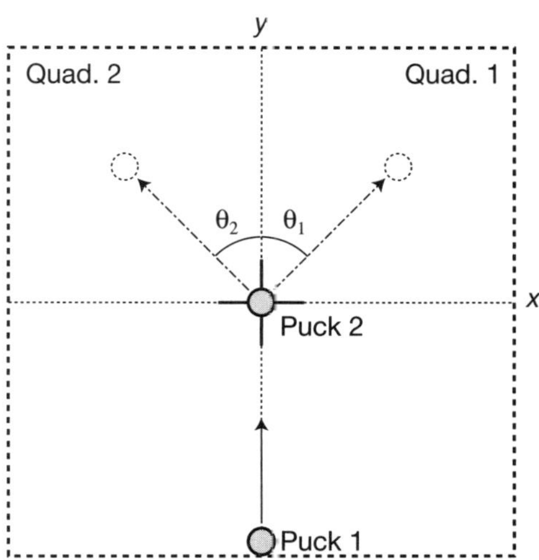

Chapter 7

Circular Motion

Spin Me Around

A ball twirled on a string is used to illustrate tangential velocity and centripetal force.

Tangential velocity • Centripetal force • Centrifugal force • Centrifugal effect

Theory

For an object to travel in a circle, a force that is continuously directed towards the center of the circular path must be applied to the object. In this demonstration a ball is swung on a string, causing it to move in a circle. Upon release of the string the center-directed force called the centripetal force, F_c, is lost resulting in the object moving off at a tangent to the circle at the point of release.

The term "centrifugal force" is often used to describe the force experienced by a passenger in a turning vehicle. A passenger in the front seat of a car "feels" as if there is a force pressing her into the door as the car makes a sharp left turn. The feeling is quite real, but the force is said to be "fictitious." (Some educators use the term "centrifugal effect" to promote a better understanding.) There is no force pushing the passenger outward. The passenger is viewing the action in a non-inertial frame of reference, whereas when viewed from the road, the passenger exists in an inertial frame of reference. Objects in an inertial frame of reference travel at a constant speed and in straight lines; that is, they do not accelerate or rotate. In an inertial frame of reference, Newton's First Law holds true. An object in a non-inertial reference frame is undergoing linear acceleration or rotation. The force experienced by the passenger is actually the inward-acting centripetal force, which redirects the passenger from her otherwise straight path. According to Newton's First Law, the passenger has inertia and therefore has a tendency to travel in a straight line. In the non-inertial frame of reference, as a car turns, it actually collides with the passenger, and pushes her towards the center of the circular turn. The push of the car door on the passenger makes her feel as if she is being kept from being thrown outward. If viewed from the outside, it is clear that if the passenger door were to fly off, the passenger would leave the car in a straight line that is tangent to the circular path as the car continues to turn. The passenger would not travel in a radially outward direction despite the feeling.

In this demonstration as the velocity of a ball increases, the force needed to keep the ball in a circular path also increases, showing that the centripetal force is proportional to the tangential velocity of the ball. In fact,

$$F_c = \frac{mv_T^2}{r}$$

<table>
<tr><td>Materials</td><td>

Needle-nose pliers

Nylon string or mason's string, 1 m

Permanent marker, black

Polystyrene ball, 4″ high-density or spherical stuffed toy

Rubber bands, #32, 3″ × ⅛″, 3

Washer, ¼″

Wire, 14- or 16-gauge, 15 cm

</td></tr>
<tr><td>Safety
Precautions</td><td>

Use caution when swinging the ball and string. Practice the swinging technique and release of the ball ahead of time to avoid hitting students. Always follow laboratory safety rules.

</td></tr>
<tr><td>Preparation</td><td>

Use needle-nose pliers to form a small loop in one end of a 15-cm piece of wire to form a device akin to a sewing needle. Thread a 5 cm section of a 1 m long nylon string through the eye of the needle. Use pliers to compress the eye to minimize the diameter of the loop. Thread the needle and string through a 4″ high-density polystyrene ball. Replace the needle by tying the string to a ¼″ washer. Tie three rubber bands end-to-end and secure the chain to the free end of the string. Use a permanent marker to color the knots between the rubber bands. As an option, a textile elastic band may be substituted for the rubber bands.

</td></tr>
<tr><td>Demonstration</td><td>

Begin by swirling the polystyrene ball in a vertical plane facing the audience. Hold the end of the string with the rubber bands dangling in your palm, as the bands are not needed for this phase of the demonstration. Ask students what the ball "feels" as it travels in a circular path. As a reference, have the students imagine that the plane of motion models the face of a clock, using 12, 3, 6 and 9 as reference points. Continue spinning the ball and then release the string when the ball is at either 3 or 9 o'clock. Depending on the direction of rotation, the ball will either move towards the ceiling or towards the floor—at a tangent to the circle. Repeat the demonstration with different release points, asking students each time to predict the trajectory of the ball. Discuss how the centripetal force, F_c, maintains the circular path of the ball, the loss of which results in the ball moving off at a right angle to F_c. Discuss the concept of inertial and non-inertial frames of reference to differentiate the concepts of centripetal force and the "fictitious" centrifugal force experienced by the ball. Perhaps relate the movement of the ball to a passenger riding in a turning vehicle.

Next, while holding the end of the rubber bands attached to the string, swirl the system above your head in a horizontal plane. Ask students to observe what happens to the rubber bands as you increase the velocity of the ball. As the ball is swung faster, the rubber bands will stretch, indicating an increase in F_c.

</td></tr>
</table>

I'm Going in a Straight Line

A small steel ball rolling in a circular track is suddenly released showing that its trajectory is tangent to the circle.

Application

Tangential velocity • Centripetal force • Normal force

Theory

In this demonstration a ball is moving inside a circular track provided by a metal hoop. According to Newton's First Law an object will stay in motion in a straight line unless acted upon by an outside force. In this case, the outside force is the normal force provided by the hoop acting on the ball. The normal force is equal to the center-seeking centripetal force that moves the ball in a circle. This force is not unlike the force of a car door pushing on a passenger, which causes the passenger to move with the car as it begins making a left turn. If the car door is suddenly removed, the passenger will continue moving in a straight line, and consequently out of the car. The ball-and-hoop assembly simulates the motion of the passenger-car system described above.

Materials

Adhesive, construction

Board, masonite or plywood, $24'' \times 24'' \times \frac{1}{4}''$

File, metal, fine, or 150-grit sand or emery paper

Hacksaw, fine-toothed

Metal macramé hoop, 6″ or 8″, ¼″ diameter

Permanent marker, black or metallic paint (depends on type of wooden board used)

Steel ball, ½″

Optional: Two different colors of spray paint

Note: An embroidery hoop may be used in place of the macramé hoop.

Safety Precautions

Although this demonstration does not pose any obvious risks, always follow laboratory safety rules.

Preparation

Use a fine-toothed hacksaw to remove a 3″ section from the hoop, making the cuts angled away from the radials of the circle, as illustrated in the diagram. Use a file to smooth the edges of all cuts. In order for the cutout section to fit into the opening of the edge of the hoop, carefully bend the larger portion of the hoop so that the fit is snug. Paint each section a different color, if desired. Mount the hoop to the center of a wooden board using a bead of construction adhesive along the outer edge of the hoop. Use a black permanent marker to mark a selection of potential space trajectories of the ball when it leaves the hoop.

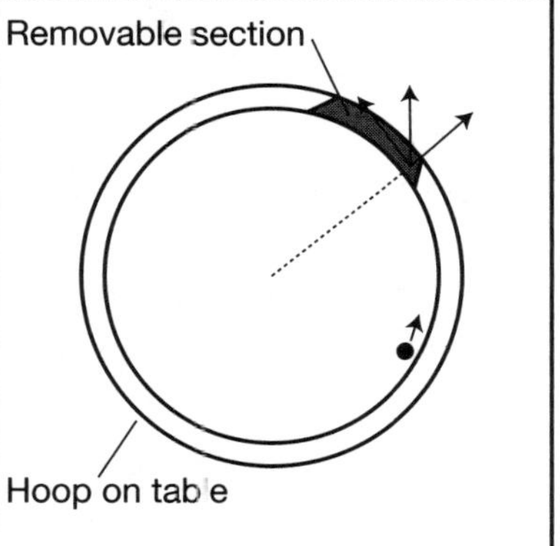

Demonstration | Place the demonstration device on a large table and invite students to stand around the table. Place the steel ball inside the hoop and launch it around the hoop in a manner similar to the way a ball spins around a roulette wheel. Ask students to predict the path of the ball if a section of the hoop is removed. After discussing the various possibilities, roll the ball and then remove the section to verify students' predictions.

A Real Swinger

A tray suspended from strings with a cup of water resting on it is swung in a circle without spilling the contents.

Application | Centripetal force • Normal force • Friction • Centrifugal force • Inertia

Theory | In this demonstration a cup of water is placed on a tray suspended from strings. The tray is twirled in a circle without spilling the contents or upsetting the cup. According to Newton's First Law, an object will stay in motion in a straight line unless acted upon by an outside force. In this case, the outside force is the normal force provided by the tray acting on the cup. The normal force is equal to the center-seeking centripetal force that moves the cup in a circle. This force is not unlike the force of a car door pushing on a passenger, which causes the passenger to move with the car as it begins making a left turn. The cup is analogous to the passenger, while the tray is comparable to the car door. In addition to the centripetal force, friction is needed to keep the cup from sliding off the tray. The frictional force, F_f, is dependent upon the normal force, N, provided by the tray, which in turn depends on the mass of the cup and its contents:

$$F_f = \mu N = \mu F_c = \mu \left(\frac{(m_{tray} + m_{cup} + m_{fluid})v_T^2}{r} \right)$$

The demonstration is more difficult to perform successfully if the cup is empty, demonstrating that the normal force and consequently friction is a major principle enabling the cup to remain on the tray. When spun in a vertical circle, the speed with which the tray is twirled also has a major effect on the success of the operation of the system. As the equation indicates, if the tray moves too slowly, then the tendency to move outward (specifically, upward) will not be sufficient to counteract the downward pull of gravity on the cup, resulting in the cup falling down. This is not unlike a passenger in a car making a left turn—the faster the turn, the greater the tendency for the passenger to continue moving in a straight line, which the passenger interprets as being thrown outward. This tendency is often referred to as a "feeling" of centrifugal force, which is not a true force. The centrifugal force is more of an idea rather than a force. (See *Spin Me Around,* page 126, for a discussion regarding centrifugal force.) In the case of the whirling tray, the inertia of the cup has a tendency to move in a straight line; however, the turning tray gets in the way and subsequently turns the cup.

As the mass of the cup increases, its centripetal force also increases. If the mass of the cup is greater than the tray, this ensures that the centripetal force acting on the cup alone is greater than the centripetal force acting on the tray alone—making the cup seem like it is being pushed into the tray even more during twirling. In fact, it is the push on the tray (normal force) that keeps pushing the cup into the circular path.

<table>
<tr><td align="right">Materials</td><td>

Drill

Drill bit, $\frac{3}{16}''$

Plastic cup, small (e.g., 8 oz)

Plexiglas®, $\frac{3}{16}'' \times 6'' \times 6''$ (available at home hardware stores)

String, nylon, $\frac{1}{8}''$ or mason's string, 4 m

</td></tr>
</table>

Materials

Drill

Drill bit, $\frac{3}{16}''$

Plastic cup, small (e.g., 8 oz)

Plexiglas®, $\frac{3}{16}'' \times 6'' \times 6''$ (available at home hardware stores)

String, nylon, $\frac{1}{8}''$ or mason's string, 4 m

Safety Precautions

Follow manufacturers' safety guidelines whenever using power tools. When twirling the apparatus, be aware of the objects around you, being careful not to collide with your body, students, ceiling fixtures, tables or other obstacles. When twirling the system in a vertical circle, twirl it so the plane of movement faces the audience. Practice the swinging technique and the demonstration beforehand, starting with slow, steady motions before attempting more dramatic swings above the head.

Preparation

Use a $\frac{3}{16}''$ drill bit to make a hole in each corner of a $6'' \times 6''$ piece of Plexiglas, $\frac{1}{2}''$ in from each side. Thread a 4-m single piece of nylon string through the holes as illustrated in the diagram, such that the "x" lies below the surface of the tray.

Tie the strings together forming a pyramid with the strings. Leave a 30-cm piece to form a loop for your hand to go through when twirling the apparatus.

Demonstration

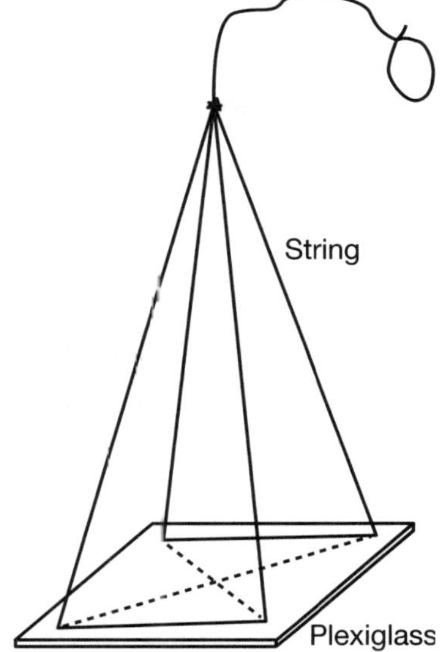

While holding a plastic cup filled halfway with water, ask students what can be done to position the cup upside down without spilling the water. Covering the opening of the cup is not allowed. Following discussion, place the cup in the center of the tray and begin swinging it gently like a pendulum taking care that it does not collide with yourself, tables or any other objects. Ask students why the cup does not slide off the tray relating the frictional forces to the centripetal force acting on the system. Continue to increase the amplitude of the system until you can continuously twirl it in a vertical circular plane. When twirling the system in a vertical circle, also twirl it so the plane of movement faces the audience. Discuss why the cup does not fall off the tray when the tray is directly above you. To stop the vertical swing, slowly decrease the amplitude of the swing by decreasing the centripetal force supplied by your hand. Finally, move about the room with the hanging tray, making sharp turns with your body and sudden stops to show how the swing of the tray compensates for every change in momentum.

Some practice is required in order to build the confidence necessary to pick up on the nuances needed to perform this demonstration successfully.

Fishing for Direction

A homemade accelerometer is used to illustrate the direction of a centripetal force.

Application

Centripetal acceleration • Newton's First Law • Centripetal force • Density

Theory

According to Newton's First Law, an object will stay in motion—in a straight line—unless acted upon by an outside force. In the case of an object moving in a circle, the centripetal force provides the centripetal acceleration, which prevents the object from continuing on its straight path. Counterintuitively, the object in this demonstration, a float in a jar of water, moves toward the center of a circle giving the appearance that is is moving inward on its own. In fact, the water and float align themselves according to their densities with the less-dense material being pushed inward by the other substance. Because of its greater inertia, the water has a greater tendency to continue in a straight line, thereby pushing the bob inward.

Consider the example of a helium balloon floating in a car. As the car accelerates, the balloon moves forward because the denser air inside the car is left behind due to its relatively greater inertia. For the same reason, the balloon moves to the back of the car when the car slows down because the relatively greater inertia of the moving air causes it to continue to move backward.

Materials

Fishing float, ¾″

Fishing line

Hot glue

Hot glue gun

Jars with lids, plastic (18-oz empty peanut butter jars work very well), 2

Lazy-Susan assembly, 12″ (available at home hardware centers)

Pendulum bob on a string

Plywood, smooth, ½″ × 24″ × 24″

Screwdriver

Velcro, self-adhesive, enough to cover both lids

Water

Optional: Band saw or jigsaw and paint or stain

Preparation

Mount the Lazy-Susan assembly in the center of a piece of plywood. As an option, the plywood may be cut into a circle using a band saw or jigsaw. It may also be painted for appearance. Tie a 20-cm piece of fishing line to a fishing float. Use hot glue to secure the free end of the fishing line to the center inside of the jar lid, adjusting the length of the line such that the float will be about 1 cm from the bottom of the jar when the lid is affixed to the jar. Build another assembly with the second jar.

Completely fill both jars with water. While working above a sink, place the float inside the jar and secure the lids to the jars. Cover the lids with self-adhesive Velcro and use it to secure the jars at opposite sides of the plywood surface. Velcro is used to enable the jars to be removed easily and emptied for storage.

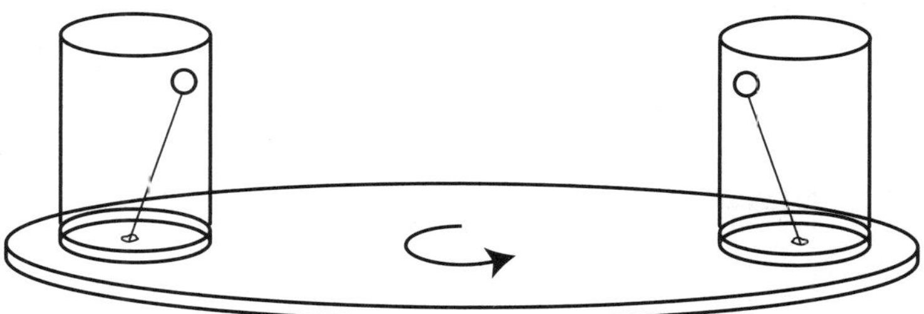

Although this demonstration does not pose any obvious risks, always follow laboratory safety rules. Clean up spills immediately.

Demonstration

Direct students' attention to the accelerometer apparatus and ask them what will happen to the fishing floats if the apparatus is spun. Following discussion, spin the apparatus gently. Ask students to explain why the floats point towards the center of the apparatus. Twirl a pendulum bob in a gentle conical circle and compare its movement to the floats in the jars. Discuss how the relative densities of the float and water play a part in the movement of the float. Ask students to explain what happens to a helium balloon in a vehicle as the vehicle accelerates or decelerates.

Spin-offs

Application | Centripetal force • Friction • Tangential velocity • Angular velocity

Theory | In this demonstration a series of identical coins arranged on a radial line is placed on a phonograph that slowly increases in speed (angular velocity). As the speed increases, the coins slide off the turntable one-by-one, beginning with the outermost coin. This illustrates the relationship between the radius of motion and the friction needed to maintain the position of each coin:

$$F_f = \mu N = \mu mg = \frac{m v_T^2}{r}$$

Although each coin has a different tangential velocity due to its position with respect to the center of the turntable, the angular velocity is the same for all coins.

The ability of an object on a turning surface to maintain its position is independent of its mass:

$$\mu mg = \frac{m v_T^2}{r} \quad \text{or} \quad \mu g = \frac{v_T^2}{r}$$

Similarly the ability of a car to turn on a surface is independent of its mass. In all cases, the coefficient of friction must be the same in comparison.

Finally, the maximum tangential velocity of an object moving in a circular path is dependent on the coefficient of friction between the two surfaces.

Materials | LP record

Phonograph (i.e., record player)

Rheostat, AC, 120-V

Optional: Overhead transparency sheet, drawing compass, scissors, marker, hole punch, fine grit sandpaper, and glue.

Note: Phonographs or turntables can be hard to come by. Consider acquiring one from the school library or media center or buying one at a garage sale. You may find a turntable an invaluable device for demonstrating circular and rotational motion. If a turntable is not available, a Lazy-Susan (available at home hardware centers) may be used in its place.

Preparation	If an LP record is not available, use a drawing compass, marker and scissors to cut a 90° pie-shaped wedge from a piece of overhead transparency that will fit on the turntable. Use a hole punch to punch a hole near the apex of the wedge.
Safety Precautions	Wear safety glasses when performing this demonstration and always follow laboratory safety rules.
Demonstration	Invite the students to gather around the demonstration table. Set the phonograph to its highest speed and plug it into the outlet of a 120-volt AC rheostat. Place an LP record on the turntable of the phonograph. Place three nickels or quarters along a radial line on the LP record. Slowly increase the output of the rheostat until the turntable begins to move smoothly. Ask students to predict what will happen if the speed of the phonograph is increased. After discussion, slowly increase the output of the rheostat to verify students' predictions. Discuss the relationship between the tangential velocity and the frictional force required to keep the coins in place. Point out that although the tangential velocity is different for each coin due to its position with respect to the center of rotation, the angular velocity is nevertheless the same for all coins. Relate the demonstration to the case of a car moving in a turn. The greater the speed of the car in the turn, the greater the frictional force required to keep the car turning in a circle. If the speed of the car is excessive, then it is possible to overcome the maximum frictional force the tires on the road may provide.

Finally, place a dime, nickel, and quarter along the edge of the record, spacing them a few centimeters apart, and repeat the demonstration to illustrate the effect of mass on the frictional force needed to maintain the position of the coins. Students will find that all the coins will slide off at the same time. This demonstrates that the ability of an object on a turning surface to maintain its position is independent of its mass. Similarly, the ability of a car turning on a surface is independent of its mass. In all cases, the coefficient of friction must be the same in comparison.

For further exploration, glue a piece of fine-grit sandpaper to the bottom of one of the coins and repeat the demonstration to illustrate the effect of the coefficient of friction on the ability of the object to maintain position as it turns.

What Makes the World Go Around

A model of a swing ride is used to explore the variables of circular motion.

Application | Amusement park rides • Conical pendulum
Tangential velocity • Angular velocity

Theory | The twirl of a ball gown, a tetherball, and a swing ride in an amusement park are all examples of a conical pendulum. In all cases, the deflection of the object is independent of the mass of the pendulum

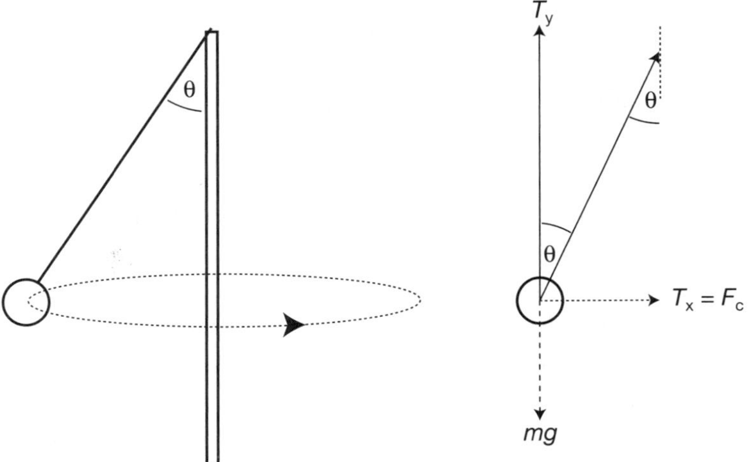

There are two forces that act on a conical pendulum: the tension, T, in the string and the force of gravity. In the y-direction the forces acting on the pendulum include gravity and the y-component of the tension:

$$\Sigma F_y = 0 = T_y - mg \quad \text{or} \quad T\cos\theta - mg = 0 \quad \text{or} \quad T = \frac{mg}{\cos\theta} \qquad (1)$$

In the x-direction, the only force acting on the pendulum is the x-component of the tension. This force serves as the centripetal force, F_c that keeps the pendulum moving in a circular path:

$$\Sigma F_x = T_x = F_c = ma_c \quad \text{or} \quad T\sin\theta = \frac{mv_T^2}{r} \qquad (2)$$

Substituting Equation (1) into (2) results in:

$$\frac{\sin\theta\, mg}{\cos\theta} = \frac{mv_T^2}{r} \quad \text{or} \quad \tan\theta = \frac{v_T^2}{gr} \qquad (3)$$

Note how the angle of deflection is dependent upon both the tangential velocity, v_T, and the radius, r, of the circular path.

Materials	Hot glue

Materials	Hot glue	Steel nut, ⅜″
	Hot glue gun	Steel nut, ⅝″
	Fly-fishing line, about 2 m	Steel nuts, ½″, 4
	Phonograph (i.e., record player)	Wire spool, empty, 6–10″ in diameter (may be procured from hardware or electronic stores)
	Scissors	
	Rheostat, AC, 120-V	
	Optional: Spray paint, 3 different colors	

Safety Precautions

Although this demonstration does not pose any obvious risks, always follow laboratory safety rules.

Preparation

Construct a device that resembles a swing ride commonly found in an amusement park. Use fly-fishing line to secure three ½″ nuts along a radial on the inside surface of one plane of the spool. Have these nuts suspended about 2–3 cm apart. The nuts should move freely without coming in contact with the plane on the opposite side of the spool. Secure a set of three different sized nuts to the opposite outside edge of the spool such that they are spaced 2–3 cm apart. As an option, consider painting each different sized nut with a unique color to differentiate them from one another.

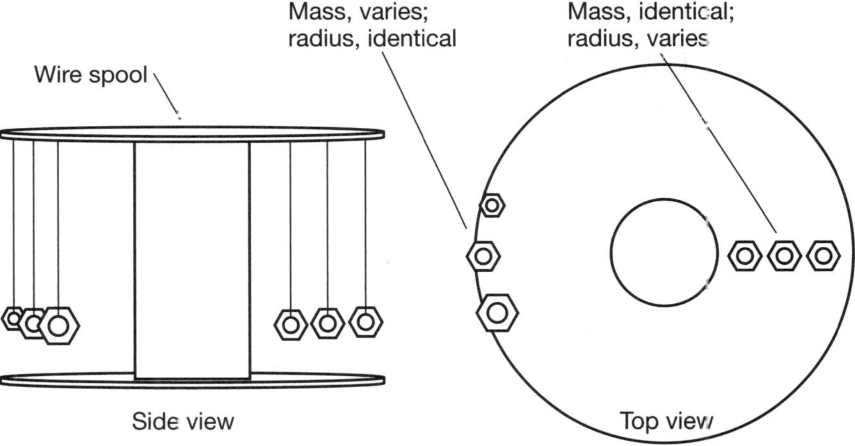

Demonstration

Set the phonograph to its highest speed and plug it into the outlet of a 120-V AC rheostat. Place the swing ride model on the turntable of the phonograph. Direct students' attention to the set of nuts located along the circumference of the spool. Point out that these nuts all have a different mass. Ask them to predict what will occur when the spool starts to rotate. After discussion, use the rheostat to slowly increase the speed of the turntable. Emphasize that the nuts along the outer circumference of the spool are all deflected the same angle despite having different masses. While the turntable continues to rotate, turn the students' attention to the other set of nuts which are lined up along a radial. Ask the students to explain why these nuts have different angles of deflection even though they all have the same mass. Finally, ask students if they were to board such a ride with a friend, would it matter where each sat.

A DEMO A DAY ———————————————————————————— **137**

Ups and Downys of Centripetal Force

A commercial laundry fabric softener–dispensing device is used to explore vertical circular motion.

Application | Vertical circular motion • Centripetal force • Newton's First Law • Torque

Theory | A Downy Ball® is a device used to automatically dispense concentrated fabric softener into the rinse cycle of a washing machine. The device works on the principle of centripetal force and Newton's First Law. As the laundry spins at the end of the wash cycle, the Downy Ball moves towards the side of the laundry drum. The ball has a spherical-like shape with its top and bottom cut flat, contributing to its function. As the laundry spins, the liquid in the ball flows to the curved side of the ball (due to the larger radius) causing the ball to orient itself with its top or bottom facing upward. The velocity of the drum increases as the water drains from the drum. When the centripetal force is sufficient, the stopper is released. This occurs when the maximum velocity is approached as the water drains. The release of the stopper may also be explained using Newton's First Law, which states that an object will stay in motion and in a straight line unless acted upon by an outside force. In the case of the Downy Ball moving in a circle, the inward push of the normal force provided by the drum provides the centripetal acceleration, which prevents the ball from continuing on its straight path. The centripetal force does not act on the stopper, which therefore has a tendency to keep moving. Consequently the stopper pops inward.

In this demonstration the Downy Ball is twirled in a vertical circular plane. Curiously, the separation of the stopper and ball occurs only at the bottom of the path indicating an unbalanced force. The two primary forces acting on the system are the centripetal force acting on the ball and the force of gravity acting on the stopper. At the bottom of the path, these two forces act in opposite directions to one another, causing the ball and stopper to separate from one another. Furthermore, the sum of these two forces must be greater than the static frictional force holding the stopper in place.

Note: The separation of the stopper from the ball may also be discussed in terms of the torque acting on the stopper and ball, since the lever-like design of the stopper assembly contributes to its release.

Materials | Downy Ball®

Drill

Drill bit, ⅜″

Eyebolt, ⅜″, 1″ long

Mason's string, 1 m

Nuts, 2

Pliers

Safety Precautions

Follow manufacturers' safety guidelines whenever using power tools. When twirling the apparatus, be aware of the objects around you so the apparatus does not collide with your body, students, ceiling fixtures, tables or other obstacles. When twirling the apparatus in a vertical circle, twirl it so the plane of movement faces the audience.

Preparation

Pry off the collar from the top of the Downy Ball. Use a $\frac{3}{8}''$ drill bit to drill a hole in the center of the Downy Ball. Use two nuts—one on the inside and the other on the outside of the Downy Ball, to secure the eyebolt through this hole. Use pliers to tighten the nuts securely. Tie a 1-m piece of mason's string to the eyebolt. On the opposite side of the string, create a loop large enough to slip a hand through.

Demonstration

Direct students' attention to the Downy Ball, and explain how it works. Seal the Downy Ball by pulling the stopper outward until it locks into position. Loop the string around your wrist and begin twirling the ball in a vertical circle so the plane of movement is facing the audience—being careful not to collide the ball with your body, students, ceiling fixtures, tables or other obstacles. Slowly increase the tangential velocity of the ball until an obvious "pop" sound is heard, indicating that the stopper has been released. Stop twirling the device and show students that the stopper is now loose. Ask students why the stopper became undone. Discuss how the inertia of the weighted stopper causes it to pop open. Repeat the demonstration a few more times, asking students to listen carefully for the pop. Ask students if they notice when the stopper pops open. The stopper is expected to pop open at the bottom of the swing. Discuss the other force acting on the stopper, gravity, which contributes to the release of the stopper. Finally, discuss how the Downy Ball may serve as a model of the forces acting on the passengers of a roller coaster ride.

Loop the Loop

A model of a roller-coaster loop is used to illustrate the conservation of energy in circular motion.

Application | Conservation of energy • Amusement park rides • Vertical circular motion

Theory | A model of a roller-coaster may be used to illustrate the law of the conservation of energy. In this demonstration a ball is released from the top of a ramp entering a loop. The motion of the ball depends upon the initial potential energy given to the ball, which is height dependent. As the ball travels along the track, its potential energy is gradually converted into kinetic energy until the ball reaches the lowest point along the track prior to entering the loop. The successful completion of the loop depends upon the velocity of the ball as it enters the loop, which in turn is dependent upon the release height of the ball. The process may be analyzed in terms of forces, energy, and the radius of the loop:

In terms of forces:

$$N + mg = \frac{mv_T^2}{r}$$

The minimum velocity required for a successful run through the loop occurs when the normal force is equal to zero:

$$mg = \frac{mv_T^2}{r} \quad \text{or} \quad g = \frac{v_T^2}{r} \quad \text{or} \quad rg = v_T^2$$

In terms of energy:

$$mgh_i = mgh_f + \frac{1}{2}mv_T^2 \quad \text{or} \quad gh_i = gh_f + \frac{1}{2}v_T^2$$

Combining the force statement with the energy statement results in:

$$gh_i = gh_f + \frac{1}{2}rg \quad \text{or} \quad h_i = h_f + \frac{1}{2}r$$

Since $h_f = 2r$ (the top of the loop), $h_i = 2.5r$, showing that the minimum release height is solely dependent upon r (assuming friction and rotational inertia of the ball are ignored, which is reasonable).

Materials	Bookshelf track, 6′ (available at home hardware centers)

Golf ball

Hacksaw, fine-toothed

Paint bucket, empty, 5-gallon

Wood glue

Wooden board, $1'' \times 3'' \times 24''$

Wooden board, $1'' \times 2'' \times 18''$

Wood screws, 2½″, 2

Wood screws, ¾″, 6

Optional: Drill and drill bit, $^1/_{16}''$

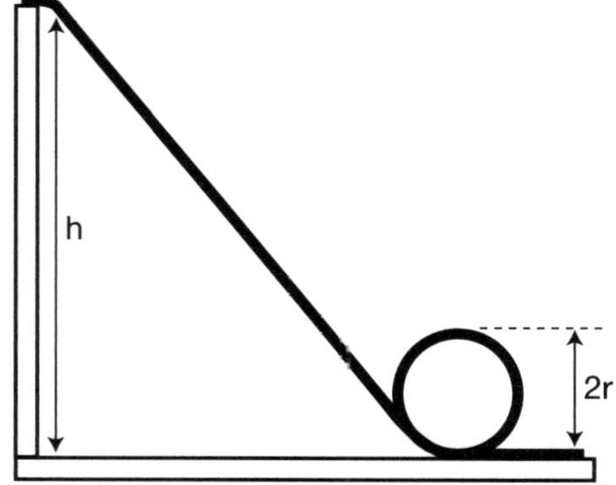

Safety Precautions

Always follow manufacturers' instruction when operating power tools. Beware of projectiles and follow laboratory safety rules.

Preparation

Attach the shorter board to one end of the longer board using wood glue and screws to form an "L" shape. Pre-drilling the holes will prevent the wood from splitting. Using a 5-gallon paint bucket as a form, bend the shelving track at one end to form a circular loop one quarter from the end of the track. Use the remaining track to make an approach ramp (see diagram). Be careful not to bend too far or the track may kink. Use the short wooden screws to secure the track to the wooden frame, bending and/or trimming the ends of the track as necessary.

Demonstration

Direct students' attention to the model of a roller-coaster loop and ask them from what minimum height must a golf ball be released to complete the loop. After discussing the various proposals, test the students' hypotheses by releasing the ball from the top of the rollercoaster. Repeat the demonstration a few times—each time lowering the release height. Show students that the minimum release height is a function of the radius of the loop.

A Shapely Planetoid

A homemade device is used to simulate the forces that lead to the oblate-spherical shape of planetoids.

Application | Centripetal force • Oblate spheroids • Acceleration due to gravity • Springs

Theory | Theories suggest that the Earth began as a molten blob of matter that coalesced into a spherical shape due to gravitational forces. If gravity were the only force acting on the planet, then the resulting shape would be a near perfect sphere. Because the Earth is also spinning on its axis, however, the centripetal force also contributed to the shaping of the Earth as it cooled, resulting in the final oblate spherical shape we are familiar with today. Nonetheless, photographs make Earth *appear* as if it is a sphere. A comparison of the acceleration due to gravity at the poles versus the equator confirms that the force of gravity is 0.34% less at the Equator than at the poles, demonstrating that the radius is also greater at the equator than at the poles:

$$g = g_0 - 4\pi^2 R/T^2 = 9.81 \; m/s^2 - 4\pi^2(6.38 \times 10^6 \; m)/(86,400s)^2 = 9.78 \; m/s^2$$

In this demonstration a pair of metal hoops is used to show how a sphere transforms into an oblate spherical shape when rotated. The change in shape demonstrates how the centripetal force acting on the system plays an important part in shaping planetoids. The spring tension is a model for the force of gravity holding the spherical shape.

Materials | Adhesive, construction

Aluminum or steel rod, ½″ × 18″

C-clamps, 2

Drill bits ⁵⁄₁₆″ and ¼″

Drill (for construction)

Drill, variable speed or similar device (for demonstration)

Lumber, scrap piece (e.g., 2″ × 4″ × 12″)

Metal strapping, ½″ × 11″, 2 pieces (used for bundling lumber; available as scrap from lumberyards or home hardware centers)

Plumber's epoxy (available from home hardware stores)

Wooden blocks made of hardwood (i.e., maple), ≈ 1″ × 1″ × ¾″, 2

Safety Precautions | The device must be constructed such that it is well balanced and will not wobble while spinning. If the device wobbles it may cause it to disintegrate and potentially harm others. Always follow manufacturers' instructions when operating power tools.

Preparation | Three $\frac{5}{16}''$ holes must be drilled into each metal strapping—one into the center and one about $\frac{1}{2}''$ in from either end. To prevent injury, secure the strapping to a scrap piece of lumber and clamp both to a sturdy surface such as a workbench. Next, drill a $\frac{1}{4}''$ hole completely through the center face of one $1'' \times 1'' \times \frac{3}{4}''$ block of wood. Drill a $\frac{1}{2}''$ deep $\frac{1}{4}''$ wide hole into the center face of the other block of wood. Attach this block to the top of the aluminum rod using construction adhesive. Allow the adhesive to set overnight. After the adhesive is thoroughly set, pass the bottom end of the rod through the center holes of both metal straps. Then, carefully bend the straps around one at a time and slip the rod through the outer holes of each strap. Finally, slide the second block over the bottom of the rod approximately $6''$. Adjust the position of the block so the straps form a spherical shape. Use about a cubic inch of plumber's epoxy on the bottom of the block to prevent it from sliding from this position.

Demonstration | Secure the demonstration device to a variable speed hand drill. Adjust the hoops so they are perpendicular to one another. Introduce the device to the students, pointing out that the straps are free to slide up and down the rod. Ask what will happen if the drill rotates the device. Activate the drill and slowly rotate the device until the spherical shape begins to form an oblate spheroid. Ask students how to increase this effect. Slowly increase the speed of the drill. Finally, ask students to describe the forces acting on the system. Discuss how the demonstration may serve as an effective model to illustrate the oblate spherical shape of the Earth and other planetoids.

Notes

Chapter 8

Rotational Motion

Hold That Torque

A device is used to illustrate how the distribution of mass affects the amount of force required to maintain static equilibrium.

Application | Torque • Static equilibrium • Moment • Couple • Lever

Theory | Torque (also called a moment or couple) causes a change in rotation of an object. The ease of opening a door depends on how torque is applied. The door acts as a lever system with the hinge acting as the fulcrum and the width of the door serving as the lever arm. It is much easier to push a door open by pushing near the outer edge farthest away from the hinge, as opposed to pushing on the door nearer the hinge. Torque depends on the force and the distance from the center of rotation from which the force is applied: $\tau = Fr\sin\theta$, where r is the distance between the applied force and the point of rotation, and θ is the angle between the force and the lever arm. Because $\sin(90°) = 1$, the maximum torque occurs when force is applied to a right angle with respect to the lever arm. In this demonstration, the torque bar serves as the lever arm. The position of the hanging mass dictates the amount of torque exerted on the handle of the system. Plumbers commonly slip a metal pipe onto the handle of a wrench to extend the lever arm of the tool. This increases the torque acting on the plumbing fixture, subsequently increasing the force on the fixture. The concept of torque originated with the work of Archimedes on levers. He said that if given a long enough lever, he could move the Earth.

Materials | Adhesive, construction

Drill

Drill bit, $\frac{1}{16}''$

Hooked mass, 1 kg

PVC cement

PVC pipe, 1″, 3′ long

PVC pipes, 1″, 6″ long, 2

PVC "T-connector"

Screw eyes, ½″, 6

Safety Precautions | Although this demonstration does not pose any obvious risks, always follow laboratory safety rules. Follow manufacturers' directions whenever working with power tools.

Preparation | Connect the three pieces of PVC pipe to the T-connector and secure the connections with PVC cement. The two 6″ PVC pipes should be 180° apart on the T-connector. Allow the cement to cure for one hour. Lay the apparatus on a table and drill a hole downward about ½″ from the end of the 3′ long arm. Drill another five holes, spaced

6″ apart beginning from the first hole. The holes must be aligned on one side of the "T" (i.e., the side facing upwards). Squeeze some construction adhesive into each of the holes and smear some adhesive into threads of the screw eyes. Insert the screw eyes into the holes in the pipe.

Demonstration | Invite a student to serve as an assistant for this demonstration. Ask her to hold the torque bar by its handle with palms facing downward. Instruct the student to hold the bar in a horizontal position and to maintain the position as you place a hooked mass on the hook nearest the handle. Ask the class if it would make a difference if the mass were placed on a hook further from the handle. After a short discussion, move the mass from hook to hook progressively away from the handle. At some point the student may not be able to maintain her grip as the torque of the system increases, and the bar will droop downward.

Reference | Carpenter Jr., D. Rae; Minnix, Richard B. *The Dick and Rae Physics Demo Notebook;* Dick and Rae: Lexington, VA, 2003; p M-614.

Where's the Middle?

The center of mass of various objects is determined.

Application | Center of mass • Center of gravity • Point mass

Theory | The center of mass (CM) is the point of average mass distribution for an object. The term center of mass is often interchanged with the term center of gravity (CG). CG is the point of location of the average weight of the object. For most small objects on Earth, CM and CG are essentially located at the same point. In tall objects such as the tallest buildings in the world, CG is located below the CM because there is a greater gravitational pull on the lower stories than on the upper stories. Nevertheless, in these buildings the difference between the location of CG and CM is only about 1 mm. If an object is constructed of a uniform solid material, then the CM will reside at its midpoint or geometrical center. The mass of an entire object will behave like a point in space, often referred to as a point mass, as if all the mass were concentrated at this point. To simplify the analysis of an object's state of motion, the object is often considered a point mass. As a point mass, however, the space that the object occupies is ignored. This can present a problem if air resistance or other factors are present to a measurable degree. In linear (translational) motion, the object may be viewed as a point mass since it is the CM that follows the path of the motion. In rotational motion, an object will rotate freely around its CM.

Materials | Banana

Cardboard, cut into an irregular shape

Coat hanger

Dowel, ¼″ × 30 cm

Drill

Drill bit, ½″

File folder

Fishing weight, small, 50 g

Masking tape

Ring stand and clamp

String, 1 m

Wooden board, 30 × 50 cm

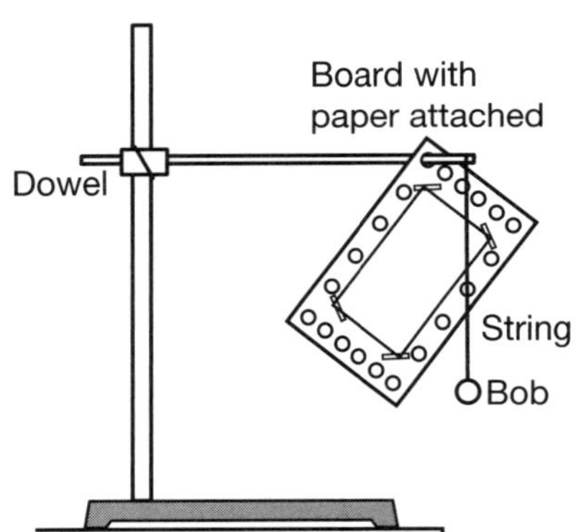

Safety Precautions | Although this demonstration does not pose any obvious risks, always follow laboratory safety rules. Follow manufacturers' directions whenever working with power tools.

<table>
<tr><td>Preparation</td><td>Drill twenty ½″ holes around the perimeter of the wooden board. The holes should be drilled completely through the board.</td></tr>
<tr><td>Demonstration</td><td>Clamp a ¼″ × 30 cm dowel to a ring stand at the edge of a lab table. Tape a sheet of paper to one side of the wooden board. Insert the board, paper side out, onto the dowel so that the board is suspended from one of its holes. Attach a plumb line (made from a string and fishing weight) to the end of the dowel. Swing the board and when it comes to rest, draw a line on the paper showing the position of the string. Repeat this procedure using two other holes to hand the board. The lines should intersect at the center of mass. Remove the dowel and board and attempt to balance the board by placing its center of mass on the dowel.</td></tr>
</table>

Use a nail or pencil to punch three holes along the perimeter of the irregularly shaped cardboard. Repeat the above procedure to locate the center of mass.

Find the center of gravity of other irregular objects. Irregular objects such as a cup, coat hanger or banana may have their centers of mass located where no actual matter exists. For some objects, including the banana and coat hanger, it will be necessary to tape the object to a sheet of paper (use a file folder) in order to draw the lines that are needed to identify the center of mass for the object.

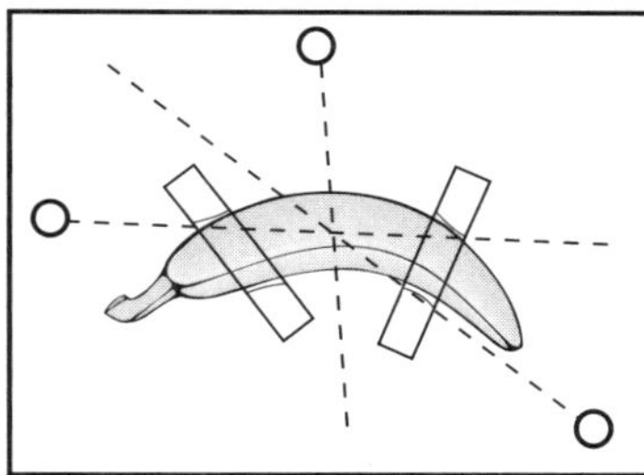 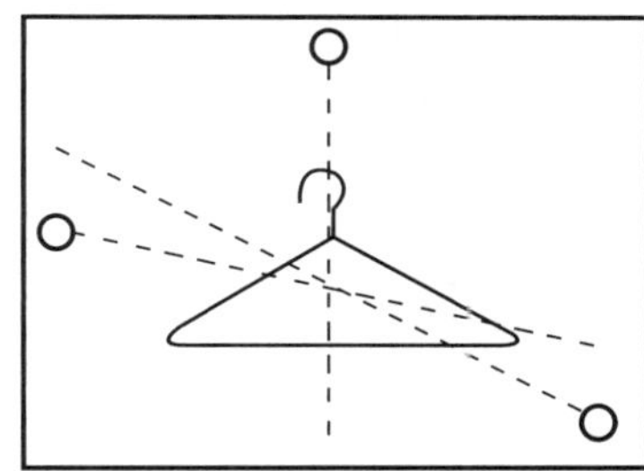

References

Bilash, Borislaw. *A Demo A Day—A Year of Physical Science Demonstrations;* Flinn Scientific: Batavia, IL, 1997; p 208. Reprinted with permission.

Faraday, Michael. *The Forces of Matter;* P. F. Collier & Son, 1910; pp 14–15.

Hewitt, Paul G. *Conceptual Physics;* Prentice Hall: Upper Saddle River, NJ, 2002, pp 138–139.

Where's the Center?

A plastic bar secretly weighted at one end is used to show that the balance point is dependent upon the distribution of mass within a body.

Application | Center of mass • Point mass • Torque • Static equilibrium

Theory | The center of mass (CM) is the point of average mass distribution for an object. If an object is constructed of a uniform solid material, then the CM will reside at its midpoint or geometrical center. The mass of an entire object will behave like a point in space, often referred to as a point mass, as if all the mass is concentrated at this point. In this demonstration, a uniformly shaped rod appears to be made of a uniform material which would suggest that its CM is located at its geometrical center. However, the object is secretly weighted on one end, which significantly shifts the CM of the rod. Consequently, the balance point will be located at an unexpected position.

In rotational motion an object will rotate freely around its CM. The rod used in this demonstration does, in fact, rotate around its CM, but the motion appears unusual because the CM is not located at the geometrical center as one would expect.

The balanced rod may also be analyzed in terms of the first and second conditions of equilibrium which state that the sum of the forces acting on the system as well as the sum of the torques acting on the system add up to zero.

Materials | Adhesive, construction

Drill

Drill bit, $\frac{1}{8}''$

Finishing nail, $3''$

Grinder, hacksaw or bolt cutters

Hammer

Lead fishing weights, 2 oz, 4

PVC end caps, $1''$, 2

PVC cement

PVC pipe, $1''$, $3'$ long

Rubber stopper, 1-hole, #00

Sandpaper, 150 grit

Tape, electrical, $\frac{1}{2}''$, lengths of at least $4''$

Wooden dowel, $\frac{1}{2}'' \times 2'$

Wooden dowel, $1''$, $6''$ long

<table>
<tr><td>Safety Precautions</td><td>Although this demonstration does not pose any obvious risks, always follow laboratory safety rules. Follow manufacturer's directions whenever working with power tools.</td></tr>
</table>

Safety Precautions

Although this demonstration does not pose any obvious risks, always follow laboratory safety rules. Follow manufacturer's directions whenever working with power tools.

Preparation

Inject about 25 mL of construction adhesive into one end of a $1'' \times 3'$ PVC pipe. Use a wooden dowel to push four 2-oz fishing weights and the adhesive into the pipe. Apply a small dollop of adhesive in between each additional weight. Use PVC cement to seal the end caps onto the pipe. Wrap a piece of ½″ electrical tape around the center of the pipe. Use a hammer to position a finishing nail into the center of one end of a short wooden dowel. Use a grinder, hacksaw or bolt cutter to cut off the end of the nail. Mount the dowel assembly horizontally to a ring stand using a utility clamp. Use a C-clamp to secure the base of the ring stand near the edge of the demonstration table so that the nail faces the audience and protrudes out beyond the edge of the table. Drill a ⅛″ hole into the geometrical center of the rod as well as a second hole in the center of mass of the rod. Be sure that the holes are in line along the length of the rod. Use a piece of sandpaper to de-burr the holes.

Demonstration

While keeping the holes in the pipe facing upwards to keep them hidden from the audience, introduce the apparatus to students and ask them what the tape in the middle of the pipe represents. After some discussion, hold the pipe by resting it on your two index fingers positioned near the end caps. Slowly move your fingers inward towards the tape until both fingers arrive near the center of mass of the pipe. Since the pipe is secretly weighted, the center of mass will not be where students will expect it to be. Ask them to account for the discrepancy.

Slide the pipe onto the nail of the dowel assembly using the center hole. Slip a #00 1-hole rubber stopper over the nail to prevent the pipe from sliding off during the demonstration. Ask students what they expect will happen if the rod is spun around on the nail. After discussing various ideas, attempt to spin the rod. It is expected that the weighted end of the rod will simply drop and barely swing, similar to a pendulum. Then, slide the pipe over the nail using the hole of the center of gravity and secure the pipe using the rubber stopper. Finally, spin the rod showing that it rotates freely when mounted on its center of mass.

Reference

Faraday, Michael. *The Forces of Matter;* P. F. Collier & Son, 1910; pp 14–17.

Mancuso, Richard V. "Quantitative Analysis of Moving Two Fingers under a Meterstick"; *The Physics Teacher,* 1993; vol 31, p 222.

Tipping Point

A plank of wood with a block resting on it is raised until the block tips over as an illustration of tipping point.

Application	Tipping point • Center of mass • Stability
Theory	When the center of mass (CM) of an object is on the same line as the pivot point of the object, the object is in static equilibrium, or balanced. In objects where the CM lies above its footprint, the pivot points are located along the edge of the footprint. If the center of mass moves outside the pivot points, the object topples over. The limit to how much an object can be tilted is known as the tipping point. A person standing will be balanced as long as her CM lies above her footprint. If a woman were to lean over to the extent that her CM is outside the vertical line above her footprint, she will tip over. When a person bends over, in order to maintain her center of gravity, she extends her hips backward in order to keep her center of mass over her footprint. In general, men are easier to tip over since their CM is higher than in women due to the fact that the average height of men is greater. In addition, men tend to be more top-heavy, which also raises the COM. The Tower of Pisa may be the most famous example of this phenomenon. Since its construction, the increase in the tower's tilt has led engineers to lower the CM so it stays above the footprint in order to prevent the tower from toppling. The CM was lowered by adding lead weights to the lower floors of the structure. In the 1990s, engineers from Rutgers University were involved in a stabilization project that attached a set of steel "suspenders" fixed to a girdle at the base of the Tower of Pisa. This process actually pulled the tower 2 cm back from its lean.
Materials	Fly-fishing line, 30 cm Mass, small, block or other small object Screw eye, ½″ Screw eye, 1″ Tape, masking Wooden block, (e.g., 2″ × 6″ × 12″ or 2″ × 4″ × 8″) Wooden board, 1″ × 4″ × 3′ *Optional:* Commercial Demonstration Model of the Tower of Pisa
Safety Precautions	Although this demonstration does not pose any obvious risks, always follow laboratory safety rules. Follow manufacturers' directions whenever working with power tools.

<table>
<tr><td>Preparation</td><td>

Locate and mark the geometrical center on the face of the small wooden block. Insert a screw eye into its geometrical center. Create a plumb bob from a piece of fly-fishing line and a 1″ screw eye. This screw eye should be suspended by its eye so the screw end is pointing downward. Suspend the plumb bob from the smaller screw eye mounted in the side of the block of wood. The length of the plumb bob should be such that the screw tip extends towards the bottom corners of the block of wood.

</td></tr>
</table>

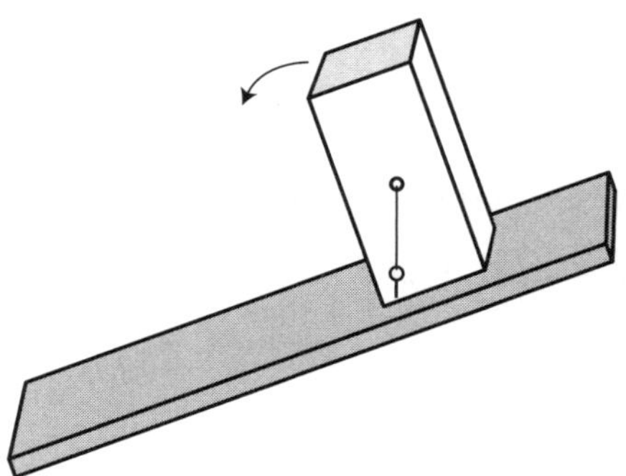

<table>
<tr><td>Demonstration</td><td>

Position the tipping block upright in the middle of the plank of wood. Ask students what will happen if the plank is lifted slowly. Test students' hypotheses by lifting the plank until the block tips over. As the plank is lifted, direct students' attention to the plumb bob and its position with respect to the base of the block. Emphasize that if the CM of the object lies above the base of the object, then the object is in rotational equilibrium and subsequently stable. If the CM lies or moves outside the base, then the edge of the base becomes a pivot point and the object begins to rotate towards the CM. Discuss how the location of the CM is taken into account when designing structures, including buildings, towers, bridges, etc.

Invite a student to assist with the second part of the demonstration. Place a 50-cm line of masking tape on the floor. Ask the student to pace out two steps in front of the tape and place an object such as a small mass or block at this spot. Have the student stand with her toes behind the tapeline and ask her to pick up the object without shifting her feet. The task should be easy to perform. Now repeat the demonstration but this time have the student stand such that her heels are in contact with a wall. No matter how hard she tries, the task should prove to be impossible to perform since she cannot keep her center of gravity above her footprint.

</td></tr>
</table>

<table>
<tr><td>Reference</td><td>

Faraday, Michael. *The Forces of Matter;* P. F. Collier & Son, 1910; pp 15–17.

</td></tr>
</table>

Balancing Act

When the center of mass of an object is located below its pivot point, it is nearly impossible to topple it over.

Application

Center of mass • Pivot point • Torque

Theory

If the center of mass lies below the pivot point, then it is virtually impossible to topple an object over since the torque exerted by the center of gravity will counter any disrupting forces. The more you try to upset the balance, the greater the effect of the torque that returns the object to rotational equilibrium.

Materials

Adhesive, construction

Coat hanger, metal

Drill

Drill bit, $1/16''$

Lacrosse or tennis ball

Solid rubber figures, bendable (e.g., Gumby® toy figure), 2

Wire cutters, heavy

Note: As an option, the concept explored here may be demonstrated using commercially made toys available from toy stores and science education suppliers. Toys include Balancing Birds, wobbling figures, or any toy that has a center of mass below its pivot point.

Safety Precautions

Although this demonstration does not pose any obvious risks, always follow laboratory safety rules. Follow manufacturers' directions whenever working with power tools.

Preparation

Drill a $1–2''$ deep hole into the body of one bendable rubber figure and into a lacrosse or tennis ball. Cut an $18''$ section of wire from a metal coat hanger. Smear a small amount of construction adhesive onto one end of the wire and insert it into the hole of the figure. Insert the other end of the wire into the lacrosse ball in a similar fashion. Apply small amounts of adhesive around the wire where it enters the figure and the ball. Allow the adhesive to set overnight. Finally, curve the wire into a semi-circular shape. Adjust the hands or feet of the figure so that it can hang from the edge of a table, fingertip or similar support.

Demonstration

Attempt to position a bendable rubber figure upright on the edge of a table. If you are able to make this figure stand, then show how easy it is to topple over. Next, place the toy on a support such as the edge of a table, the tip of a finger or similar surface. Move the figure such that it will sway back and fourth. Ask students why this figure does not fall. Students will probably realize that the ball attached below the figure has something to do with stabilizing the system. Ask students to locate the center of mass of the system and ask them how its relationship to the pivot contributes to the stability of the system.

Reference

Faraday, Michael. *The Forces of Matter;* P. F. Collier & Son, 1910; pp 14–17.

Turner, Ray; "100 Years of Physics and Toys: Balancing Toys"; *The Physics Teacher;* vol 30, p 542.

Seesaw Carts

A pair of stationary dynamics carts are forced apart by a spring (on a track balanced like a seesaw) to illustrate a dynamic system in static rotational equilibrium.

Application

Static equilibrium • Rotational equilibrium
Torque • Conservation of momentum

Theory

According to the first and second conditions of equilibrium, two or more objects are said to be in static rotational equilibrium when the sum of the forces and the sum of the torques acting on the lever arm are equal to zero. In this demonstration two equal-mass dynamic carts resting in the center of a track are balanced on a pivot. The system is in static equilibrium because the center of gravity is located over the pivot. When the carts are pushed apart by a spring trigger, the center of mass does not move because the movement of the carts is subject to the conservation of momentum. Since the distance between each cart and the pivot is the same through-out their movement, the sum of the torques applied by each car on the lever (ramp) is always equal to zero.

Materials

Balance or spring scale

Hammer, small or similar weight object

Dynamics carts, pair, one with spring

Track with magnetic end bumpers

Wooden block, squarely cut, $2'' \times 4'' \times 3''$

Note: An air track system may be used with similar results.

Safety Precautions

Although this demonstration does not pose any obvious risks, always follow laboratory safety rules.

Demonstration

Use a balance or spring scale to show students that the carts are equivalent in mass. Balance a dynamics track on a wooden block to simulate a seesaw. Cock the spring in the spring-loaded cart and position it facing the other cart on the center of the dynamics track. The carts should be in contact with one another. Ask students what will happen to each cart if the spring is released. After discussing the various ideas, use a small hammer or similar device to release the spring. The carts should move apart with equal velocities and arrive at the ends of the tracks at the same time—bouncing off the magnetic bumpers and returning to their initial position. As the cars travel toward the magnetic bumpers and back to their initial position, the track will remained balanced. Ask students why this is the case. Discuss how the center of mass of the two-cart system does not change despite the fact that the carts move apart and that the torque exerted by one cart is always balanced by the other.

Reference

This demonstration was performed by Vacek Miglus of Wesleyan University at the Demonstration Workshop during the AAPT Summer Meeting held at the University of Wisconsin, Madison, August 2003.

A device proves that objects can fall faster than "*g*."

Application | Inertia • Newton's First Law • Center of gravity • Rotation • Stress

Theory | This demonstration primarily illustrates two concepts: Newton's First Law as well as an object's speed and acceleration during rotation. A device is constructed in such a way that when the arm falls, the cup ends up in the position where the ball comes to rest. The ball follows a straight-line path as gravity pulls it downward. For the ball to land in the cup, the cup must accelerate faster than the ball. Since the ball is permitted to fall freely, it is assumed that it must be falling at a rate of acceleration equal to "*g*," 9.8 m/s^2. The arm, to which the cup is attached, is also permitted to fall freely. Most people think that all objects fall at a rate of *g*. Not true! In the case of a falling tree, the center of gravity of the tree falls nearly 9.8 m/s^2, however, the top of the tree accelerates at a higher rate because it must travel further as the tree rotates during its fall. The same applies to the ball-in-the-cup apparatus. The center of gravity of the arm falls at a rate of *g*, however, the end to which the cup is attached falls at a faster rate than *g*, thereby arriving at the base of the apparatus ahead of the ball, which falls at a rate of *g*. This phenomenon explains why falling chimneys often break in mid air as they fall. As the chimney falls, there is a lateral stress that arises during the rotation of the chimney that causes the mortar holding the bricks together to fail. Mortar withstands compression well but often fails under lateral stress.

Materials | Ball (i.e., marble)

C-clamp

Carpet tacks, ½ ", 5–10

Golf tee

Hammer

Hinge, 2″ and screws

Paper cup, small (e.g., 4-oz)

Velcro® strip, 12″

Wooden board, 1″ × 2″ × 30″

Wooden board, 1″ × 2″ × 36″

Wooden dowel, ⅜″ or ½″, 16″

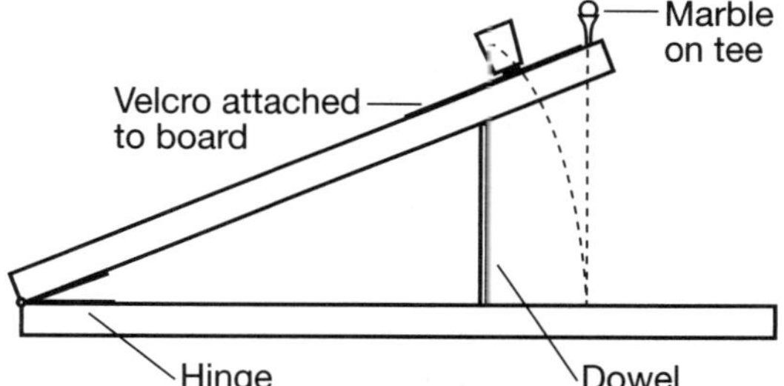

Safety Precautions | Although this demonstration does not pose any obvious risks, always follow laboratory safety rules.

Preparation	Construct the ball-in-the-cup apparatus as in the diagram. Use the shorter piece of wood for the arm. Use ½″ carpet tacks to secure a 12″ strip of Velcro to the end of the top side of the shorter piece of wood. Attach Velcro to the underside of the cup. Velcro will permit you to adjust the position of the cup.
Demonstration	Use a C-clamp to secure the base of the apparatus to a table. Balance the top arm of the apparatus on the dowel. Place the marble on the golf tee. Bring students' attention to the marble on the device. Ask students what would happen if you quickly removed the dowel from the apparatus. Ask them to predict exactly where the marble will fall. Invite one student to place the Velcro-bottomed cup on the top arm of the device in such a position that the ball will land in the cup. While holding the base of the apparatus, use your other hand to knock the dowel out of position. Ideally, the ball should land in the cup. Finally, ask students to explain the paradox of why the cup and the ball did not fall at the same rate.

Note: The motion of the objects in this demonstration is rather fast resulting in a limited time for observation. Consider videotaping the demonstration and reviewing the tape in slow motion or frame by frame.

References

Haber-Schaim, Uri. "On Qualitative Problems"; *The Physics Teacher,* 1992; vol 30, p 260.

Phelps III, F. M., Clifford, L. R. "How the Ant Got Into the Dish"; *The Physics Teacher;* 1986; vol 24, p 293.

Sutton, Richard M. "M-206: Chimney Falling — Free Fall Paradox"; *Demonstration Experiments in Physics.* McGraw-Hill: New York, 1938; p 89.

Throwing Off Your Center of Mass

The rotation of an object around its center of mass is explored.

Application Center of mass • Point mass

Theory In rotational motion an object will rotate freely around its center of mass (CM). The CM is the point of average mass distribution for an object. If an object is constructed of a uniform solid material, then the CM will reside at its midpoint or geometrical center. The mass of an entire object will behave like a point in space, often referred to as a point mass, as if all the mass were concentrated at this point. In this demonstration a uniformly shaped disc appears to be made of a uniform material that suggests its CM is located at its geometrical center. However, the object is secretly weighted on one side, thereby shifting the CM of the disc significantly. Consequently, the disc does not spin around its geometrical center. On one side of the disc the geometrical center is marked with a small target while on the other side the CM is marked with a similar looking target. The disc is painted red to make the black and white target more pronounced. Red is used because the human eye is less sensitive to it so this brings attention to the target. The second half of the demonstration utilizes a compound system of two attached balls. Because the mass of the balls varies, the CM is offset from the geometrical center of the system. Tossing the system makes it clear that the CM is outside both balls. Objects such as a boomerang, a carpenter's square, and a set of headphones have their CM located outside the object

Materials Adhesive, construction
Cardboard, ¼″ thick, corrugated, 18″ round, 3
Cardboard, ¼″ thick, corrugated, 2″ round
Computer word processor with drawing tool and printer
Drill
Drill bit, ½″
Paper clip
Razor knife
Soccer ball
Spray paint, neon green or yellow
Spray paint, red
String, woven nylon, ⅛″, about 2 m
Wooden dowel, ½″ × 6″
Wooden block, 2″ × 4″ × 4″
Slotted mass, 100 g or lead shot
Tennis ball
White glue

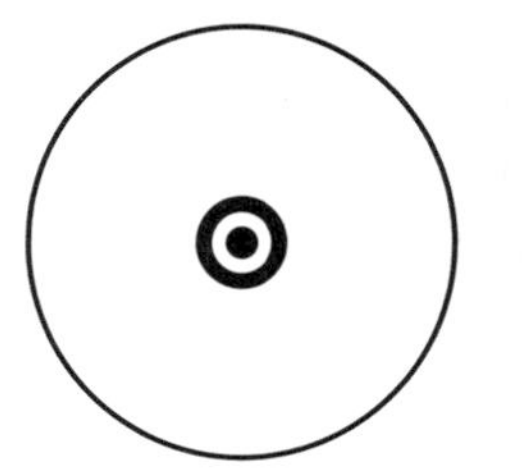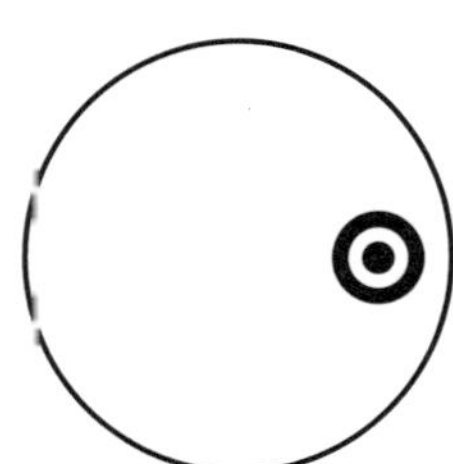

<table>
<tr><td>Safety
Precautions</td><td>Follow manufacturers' directions whenever working with power tools. Although this demonstration does not pose any obvious risks, always follow laboratory safety rules. When tossing the cardboard assembly, take caution not to throw it towards ceiling fixtures or other obstructions.</td></tr>
<tr><td>Preparation</td><td>Drill a ½″ hole 1″ into the center of the 4″ × 4″ face of the wooden block. Use construction adhesive to secure the dowel in the hole.</td></tr>
</table>

Place a slotted 100 g mass about 2″ from the edge one of the large cardboard circles. Trace an outline of the mass and then use a razor knife to cut out the hole. Use construction adhesive to sandwich this cardboard circle to a second, hole-free circle. Place a dollop of adhesive inside the hole in the cardboard and press the mass into the hole. (Lead shot may be substituted for the 100 g mass. Cut out a circle about 4 cm in diameter and fill the circle with lead shot. Use an excess amount of adhesive to secure the individual pieces of shot.) Cover this side of the double cardboard circle with a third circle of cardboard using adhesive to secure the pieces permanently to one another. Allow the adhesive to cure overnight. Paint both sides of the disc with red paint. Allow to dry thoroughly. Use the drawing tool to create two black and white 2″ diameter targets. On one side of the disc, use white glue to attach one of the targets to its geometrical center. On the reverse side, determine and mark the location of the center of mass of the disc by balancing it on the dowel stand. Use white glue to attach the other target to the center of mass of the disc.

Paint both sides of the small 2″ cardboard circle with neon green or yellow paint. Allow to dry thoroughly.

Tie a soccer ball to the end of a 2-m piece of woven nylon string. Tie the ball in the same fashion as tying a ribbon around a gift box but using six loops instead of two. Leaving about 50 cm of string between each ball, tie the opposite end of the string to a tennis ball in the same fashion. Place a paper clip on the string a few centimeters away from the soccer ball. While holding the tennis ball with the soccer ball hanging downward, toss the system in the air such that the two balls rotate around one another. Watch the position of the paper clip as the system rotates. Then move the clip toward the CM and keep tossing and adjusting until the CM is identified. Use construction adhesive to secure the small cardboard disc to the CM of the system.

Demonstration

Present the cardboard disc with the geometrically centered target facing the audience and ask students what will happen if the disc is tossed upward and spun around. After discussing the various ideas, toss the disc being careful not to reveal the backside of the disc. The audience should see the target unexpectedly spiral upward and down. Ask students if they can explain what they just observed. Discuss how a spinning object will rotate freely around its CM. Finally, present the other side of the disc and toss it in the same manner as was done previously. The audience should see the target moving without a wobble, as expected.

While holding the tennis ball with the soccer ball hanging downward, toss the system in the air such that the two balls will rotate around one another. Discuss the location of the CM and how a rotating system behaves.

References

Carpenter Jr., D. Rae; Minnix, Richard B. *The Dick and Rae Physics Demo Notebook;* Dick and Rae: Lexington, VA, 2003; p M-662.

Eckroth, Charles."Earth and Moon Motion Around their Common Center of Mass;" *The Physics Teacher;* vol 28, p 425.

Inertial Wands

Two rods with different moments of inertia are used to demonstrate the relationship between torque and rotational inertia.

Application | Rotational inertia • Moment of inertia • Torque • Center of mass

Theory | Two objects with the same mass do not necessarily behave the same despite having the same center of mass. This difference may be illustrated using two rods of identical mass but with the mass distributed differently along their lengths. One rod has most of its mass located near the geometrical center of the rod while the other rod has most of its mass located near the ends of the rod. In both cases the center of mass is located at the geometrical center of the rod. Just like a force is required to change the linear motion of an object, a torque is required to change the rotational motion of an object. If the rods are rotated while holding them at their centers, the rod with most of its mass near the center is easier to rotate than the other. The rod with its mass further from the center has a greater rotational inertia which requires greater torque to initiate rotation. After the rods are rotating at equal rates, when left alone, the rod with the greater rotational inertia will tend to rotate longer. As a figure skater spins, she brings her hands inwards, thereby decreasing her rotational inertia, which results in an increase of her rotational velocity because angular momentum must be conserved.

Materials | Adhesive, construction

Dowel, 2′, ⅛″ – ¾″

Fishing weights, 2 oz, 12

PVC end caps, 1″, 4

PVC cement

PCV pipe, 1″ × 3′, 2 lengths

Screw eyes, ½″ or cup hooks, 2

Spring scale, 5 N, 500 g

Tape, electrical, 2 colors

Safety Precautions | Although this demonstration does not pose any obvious risks, always follow laboratory safety rules.

Preparation | Inject about 10 mL of construction adhesive into one end of a 1″ × 3′ PVC pipe. Use a wooden dowel to push six 2-oz fishing weights and the adhesive toward the center of the pipe. Use the dowel to verify that the weights are equidistant from either end of the pipe. Use PVC cement to seal the end caps onto the pipe.

Inject 5 mL of construction adhesive into each end of a 1″ × 3′ PVC pipe. Use a wooden dowel to push three 2-oz fishing weights and the adhesive barely into each end of the pipe. These weights should be as far from the center as possible. Use PVC cement to seal the end caps onto the pipe. Wrap the ends of each rod with different colors of electrical tape. Attach screw eyes to the ends of each rod to serve as hooks for measuring mass and also for ease of storage.

Demonstration

Invite a student to serve as a volunteer. Provide the student with the pair of rods and ask her to hold each rod from her center and rotate the rods about the center as a majorette twirls a baton. Remind the student to keep a firm grip on the rods so that they do not slip from her hands. Ask the student if there is any difference between the rods. The class may also wish to offer their observations. The student should discern that one rod is easier to twirl, while the class may notice that one rod moves with a higher rotational velocity. Ask students why they think the rods behave differently. Students may propose that the difference is due to the mass of each rod. In response to this hypothesis, use a spring scale to measure the mass of each rod to show that the masses are the same. Use the example of a figure skater to illustrate how the distribution of an object's mass affects the rotational velocity of the object.

Race of Shape

A series of objects is released down a ramp to illustrate that their speed depends on their shape.

Application | Rotational inertia • Moment of inertia • Conservation of energy

Theory | An object positioned at the top of a ramp has potential energy, *mgh* due to its position with respect to the bottom of the ramp. As the object moves down the ramp, the potential energy is converted into kinetic energy. If the object rolls down the ramp, some of the original potential energy is converted to rotational kinetic energy:

$$mgh = \frac{1}{2}mv^2 + \frac{1}{2}I\omega^2$$

where I represents the rotational inertia of the object and ω is the angular velocity of the object. The amount of energy converted to rotational energy depends on the shape of the object. In this demonstration four objects are rolled down an incline and finish in the following order from fastest to slowest—sphere, solid cylinder, hollow cylinder, and thin hoop. The rotational inertia of each object is:

$$\text{Sphere:} \quad I = \frac{2}{5}mr^2$$

$$\text{Solid cylinder:} \quad I = \frac{2}{5}mr^2$$

$$\text{Hollow cylinder:} \quad I = \frac{2}{5}m(r_{ID}^2 + r_{OD}^2)$$

$$\text{Thin hoop:} \quad I = mr^2$$

Objects with nonuniform composition or ones in which the distribution of mass changes as they roll will exhibit chaotic behavior. Dramatic examples include cans of soup, corn niblets, and other containers holding viscous materials.

Materials | Books, several

Can of soup or corn niblets

Cylinder, solid (e.g., D-cell battery)

Hoop, thin, made of a ½″ piece cut of 2″ PVC pipe

PVC pipe, 2″, 6″ long

Sphere (e.g., golf ball)

Wooden plank, smooth, or comparable ramp (Example: 1″ × 6″ × 8′)

Safety Precautions | Although this demonstration does not pose any obvious risks, always follow laboratory safety rules.

Demonstration | Position the ramp on a stack of books on a demonstration table or floor. Explain to students that you are about to race four objects down the ramp: a thin hoop, a hollow cylinder, a solid cylinder, and a sphere. Ask students to predict the order in which the objects will finish the race. Arrange the objects one behind the other in order from fastest to slowest, with the fastest at the head of the line. If the order is correct, the objects will separate from one another as they travel down the ramp. If the prediction is incorrect, the objects will bunch up. In this case, change the order of the objects and repeat the race until the correct order is determined. Show students how knowing the moment of inertia of each object can be used to predict the order of finish.

Finally, roll a can of corn niblets down the ramp and ask students to explain the odd behavior of its movement down the ramp.

Reference | Giancoli, Douglas C. *Physics,* 5th ed., Prentice Hall: Upper Saddle River, NJ, 1998; p 226.

Spin 'em Up

A student sitting on a rotating stool spins around with and without her hands extended to illustrate how angular velocity depends on the distribution of mass.

Application | Angular velocity • Angular momentum • Moment of inertia

Theory | The angular velocity of a spinning object is dependent upon the distribution of its mass. This principle is employed in figure skating. As the skater draws her arms inward, she reduces her rotational inertia causing an increase in her angular velocity. Her angular velocity increases because angular momentum is conserved. Notice how the angular momentum is directly related to the angular velocity: $L = I\omega$, where L represents the angular momentum, I is the moment of inertia, and ω is angular velocity. The skater has a larger moment of inertia when her arms are extended than when they are pulled inwards. Since the angular momentum is constant, as I decreases, ω must increase proportionally. Competitive divers also employ this principle by tucking in their bodies as they somersault during the dive. Since the tucked shape has the least rotational inertia, it gives the most control over rotational velocity. A pair of spinning masses called a governor is used to regulate the speed of an engine. As its speed increases, the centrifugal force generated by the engine moves the spinning balls outward triggering a reduction in the throttle of the engine.

Materials | Dumbbell weights, 3–5 lb

Rotating platform

Stool

Optional: Drill and drill bit; 12″ disks cut from ¾″ plywood, 2; 12″ Lazy Susan kit, (available from hardware centers or cabinet maker supply store), and a screwdriver.

Note: A rotating platform may be fashioned from two pieces of plywood and a Lazy Susan for building a rotating platform. See the *Preparation* instructions below.

Safety Precautions | Perform this demonstration in an open area away from obstacles such as desks and chairs. Exercise care in selecting volunteers as spinning motion can cause nausea in some people. Always follow laboratory safety rules while performing demonstrations.

Preparation | Use a screwdriver to attach the plywood disks to either side of the Lazy Susan. Pre-drill the holes for the screws to ease attachment. The wood may be sanded and painted if desired.

Demonstration | Place the rotating platform on a stool. Ask a student volunteer to sit on the platform. A small-stature student will provide a more dramatic exhibit. Hand the volunteer a pair of dumbbell weights and ask the student to grasp the weights securely with arms and hands fully extended sideways. Have the student also cross his ankles and extend his feet outward slightly away from the floor and the base of the stool. Explain to the student how the demonstration will proceed and that he needs to keep his body centered over the stool. Spin the student slowly by applying a turning force to his shoulders and step back so that his arms do not collide with you. Once the student begins spinning, ask him to bring the dumbbells in towards his chest. As the dumbbells are drawn inward, the student's angular velocity will dramatically increase. After the initial trial, repeat the demonstration a few times and have the student move the dumbbells outward in order to slow the motion. Discuss how the position of the dumbbells affect the angular velocity of the student. Explain how figure skaters and high divers use this principle to contribute to their performance.

Chapter 9

Oscillations

Simple Harmonic Motion — It's Easy as π

Circular, mass–spring and pendulum motion are demonstrated together to show that each is an example of simple harmonic motion.

Application	Simple harmonic motion • Mass–spring systems • Pendulum • Circular motion
Theory	The period of a record player is adjustable—typically to three settings: 45 rpm, 33 rpm and 78 rpm. The 33-rpm setting corresponds to a period of 1.8 s. The period of a mass–spring system in simple harmonic motion may be determined using the formula:

$$T = 2\pi \sqrt{\frac{m}{k}}$$

where m represents the mass connected to the spring and k equals the spring constant. The period of a pendulum, T, is summarized by:

$$T = 2\pi \sqrt{\frac{l}{g}}$$

where l is the length of the pendulum and g equals the acceleration due to gravity, 9.8 m/s^2. In this demonstration all three devices—a record player, pendulum, and mass–spring—will be adjusted so that their periods are all the same. A period of two seconds may be achieved by using a 10 N/m spring with a 1-kg mass, and a 1-m long pendulum. The record player may be set at 45 rpm and adjusted down using a rheostat to match the periods of the spring system and the pendulum.

Materials

Dowel, ½″ d × 6″ l

LP record or cardboard disk

Mass, 1 kg

Overhead projector and screen

Pendulum bob

Pendulum clamps, 2

Record player (turntable)

Rheostat or a ceiling fan dimmer switch wired to an extension cord

Ring stands, tall, 2

Spring, $k = 10$ N/m

String, 1.5 m

Velcro® tabs, ½″ self-adhesive, 1 pair

Safety Precautions

Although this demonstration does not pose any obvious risks, always follow laboratory safety rules.

Preparation	Attach a Velcro tab to the top edge of an LP record. Place the matching Velcro tab on one end of a ½″ d × 6″ l dowel. Affix the dowel to the LP record. The dowel should protrude from the surface of the record. Place the record on a record player and situate it between a projection screen and the overhead projector. Adjust the overhead so a shadow of the dowel will be visible on the screen. Plug the record player into a rheostat and set the speed to 45 rpm. Suspend a pendulum bob from a string attached to a pendulum clamp. The pendulum system should be 1 m long. Suspend a 1-kg mass from a 10 N/m spring attached to a pendulum clamp. Set the mass–spring system in motion and adjust the length of the pendulum string to match the period of the mass–spring system. Use the rheostat to adjust the speed of the record player to match the period of the mass–spring system.
Demonstration	Begin the demonstration by setting the mass–spring system in motion. Ask students to describe the motion of the system, including the velocity of the mass at any given moment. With the mass–spring system still in motion, lift the pendulum and allow it to oscillate back and forth. Ask students to describe the motion of the pendulum including its velocity at any given position. Ask students to compare the motion of the pendulum to that of the mass–spring system. Discuss the similarity between the movement of the pendulum and the mass–spring system. Finally, turn on the turntable and ask students to describe the motion of the dowel, including its velocity at any given position. Ask students to compare the motion of the turntable to the mass–spring system. Some students may have a difficult time seeing the connection between the oscillation of a pendulum and the moving dowel. To illustrate this concept, use an overhead projector to cast a shadow of the dowel onto a projector as the dowel moves—the motion of the dowel in only one dimension will be observed.

How Does It Swing?

The variables that affect the periodic motion of a simple pendulum are explored.

Application Pendulum • Simple harmonic motion

Theory A pendulum is a body that swings freely from a fixed point in a vertical plane under the influence of gravity. A simple pendulum consists of a bob supported by a thin wire or string. When raised and then released, the pendulum bob will move from the release point through its rest position, up to a height equal to its release height, and then back to its release position. This path is called a cycle. The time taken for one complete cycle of the pendulum is called the period. The period, *T*, of a pendulum is dependent upon the length of the pendulum as measured from its point of support to the center of gravity of the pendulum (in this case, the center of the bob). The period, however, is independent of mass:

$$T = 2\pi \sqrt{\frac{l}{g}}$$

As the length of the pendulum increases, so does its period. An object is said to undergo simple harmonic motion when it moves in one direction and then returns to its starting point.

Materials C-clamps, 2

Drill

Drill bit, $\frac{1}{8}''$

Ring stands, 2

Stopwatch

String or monofilament fishing line, 6 m

Washers, steel, $\frac{5}{8}''$, 24

Wooden dowel, $\frac{5}{8}'' \times 36''$

Safety Precautions Although this demonstration does not pose any obvious risks, always follow laboratory safety rules. Always follow manufacturers' instruction when operating power tools.

Preparation Set-up five pendulums with the following configurations:

Pendulum	No. of Washers	Length of String
1	3	25 cm
2	3	50 cm
3	3	75 cm
4	6	75 cm
5	9	75 cm

A bob is made by tying three, six or nine steel washers to a length of string or thread. Drill five holes into the dowel about 15 cm apart. Attach the pendulums to the dowel as illustrated in the diagram using permanent knots. In particular, pendulums 3, 4, and 5 must be the same length. Arrange the ring stands as illustrated in the diagram. Secure the system to a demonstration table using two C-clamps.

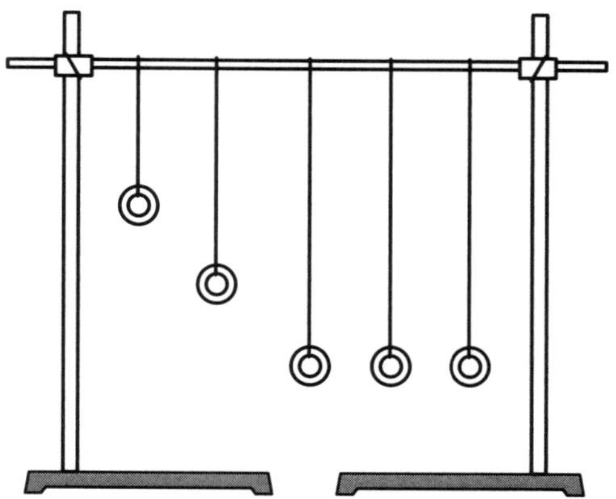

Demonstration

Swing each of the pendulums by raising and then releasing the bobs from the same height (about 15° from its rest point). Compare the period of each pendulum by counting the time taken for 10 complete cycles. Discuss what factors determine the periods of the pendulums.

A Couple of Oscillations

Six pendulums of varying and common lengths are coupled together and swung to illustrate the factors effecting resonance.

Application

Resonance • Coupled pendulums • Natural frequency

Theory

Harmonic oscillators, such as pendulums or mass–spring systems that are coupled together, have a tendency to synchronize if their lengths are identical. In the case of a mass–spring system, the mass and spring constant must be identical. The period, T, of a pendulum is summarized by:

$$T = 2\pi \sqrt{\frac{l}{g}}$$

where l is the length of the pendulum and g equals the acceleration due to gravity, 9.8 m/s^2. Given sufficient time, two pendulum clocks mounted on the same wall will eventually synchronize in their movement. When marching across a bridge, soldiers marching in cadence must break into a route step march, a normal step march, to prevent a coupled resonance with the bridge. If the cadence march were to match the natural frequency of the bridge, then the compounded vibrations could jeopardize the bridge's integrity.

Materials

Aluminum rod, ½″ d × 48″

C-clamps, 2

Drill

Drill bit, ⅛″

Golf balls or pendulum bobs, 6

Mason's string, 4–5 m

Right-angle clamps, 2

Ring stands, large, 2

Safety Precautions

Although this demonstration does not pose any obvious risks, always follow laboratory safety rules. Always follow manufacturers' instruction when operating power tools.

Preparation

Drill a ⅛″ hole through the center of each golf ball. Drill two ⅛″ holes through the aluminum rod, 6″ from each end. Thread a piece of mason's string through both holes of the aluminum rod and allow the string to drape about 5 cm as illustrated in the diagram. Tie-off the ends of the strings. Prepare three pairs of pendulums having lengths of 30, 50, and 70 cm. Attach the pendulums to the string–bar system using permanent knots, as illustrated in the diagram. Arrange the ring stands as illustrated in the diagram. Secure the system to a demonstration table using two C-clamps.

Ask students what they think will happen if one of the pendulums is raised and released. After discussing the various ideas, lift and release the shortest pendulum and observe the motion of the system for at least one minute. Repeat the procedure with the other pendulum pairs. Finally, relate the observations to examples of resonance in everyday life.

Stressed Out Stick

A model is used to illustrate how tension and compression forces act on a body.

Applications | Tension • Compression

Theory | An object that is being stretched or pulled apart is said to be under tension. An object that is squashed or pushed inward is said to be under compression. Tension and compression are concepts that are important in material science, mechanical engineering, and structural engineering. For example, in designing a building an engineer must consider the properties of the materials used in deciding how to distribute the tension and compression forces within the structure. Concrete, for example, can withstand considerable compression but fails quickly when under tension. The tensile strength of concrete is about 10 percent of its compressive strength. Steel is an excellent building material for it can withstand considerable tension and compressive forces as well. In this demonstration a foam model of a horizontal beam is analyzed as it is deformed, representing the addition of a load to the beam. When the beam is suspended from either end, with a load placed in the middle to model a simple bridge, the top layer of the beam undergoes compression, while the bottom layer undergoes tension. The middle or neutral layer experiences little of these forces. The design of an I-beam takes advantage of this fact. In an I-beam the forces are concentrated in the top and bottom flanges, leaving the middle relatively stress-free. An I-beam can withstand nearly the same load as a solid rectangular bar of the same dimensions, but it weighs considerably less because it uses less material.

Materials | C-clamps, 2

Cushion or 2″ thick piece of foam, 1′ × 1′

Hooked mass set

Polyurethane foam, oblong piece (e.g., 5-cm × 10-cm × 50-cm)

Permanent marker

Right angle clamps, 2

Ring stand, 2

Ruler

Wooden dowel, ⅛″ d × 36″ (one per demonstration)

Safety Precautions | Although this demonstration does not pose any obvious risks, always follow laboratory safety rules.

Use a permanent marker to bisect the length of the 10-cm (W) × 50-cm (L) face of the foam. Draw a series of lines perpendicular to the first line every 3 cm.

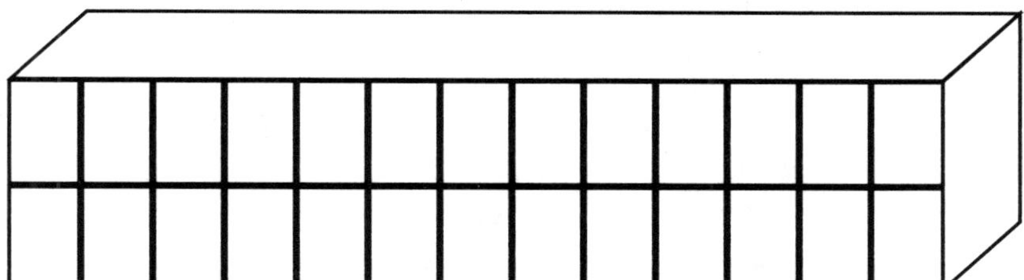

Suspend a ⅛″ × 36″ wooden dowel horizontally from two ring stands using right angle clamps. The dowel should be about 50 cm above the table surface. Use C-clamps to secure the stands to the demonstration table. Place a piece of foam or a pillow under the apparatus to catch falling weights.

Demonstration

Direct students' attention to the horizontally-mounted wooden dowel and propose that the dowel represents a model of a beam. Ask students what will happen to the dowel if a weight is suspended from its middle. After a short discussion, hang a 1-kg mass from the dowel. Students should notice that the dowel will bend downward. Discuss the compression and tension forces acting on the dowel. Continue adding mass until the dowel snaps.

Hold the foam model of the beam from either end and compress it so it takes the shape of an inverted arch—similar to the bent dowel. Show students how the top layer is compressed, while the bottom layer is stretched.

Damp the Oscillations

A mass–spring system is set to oscillate in air and while immersed in a liquid.

Application

Damped oscillations • Resonance • Shock absorbers

Theory

Most vibrating systems analyzed in a typical introductory physics class represent an ideal system in which the oscillations occur indefinitely under the action of a linear restoring force. In reality, no such system exists since retarding forces such as friction reduce the mechanical motion of the system. Such a system is said to be damped.

The design of an automobile shock absorber is based upon the dampening of a mass–spring system. The shock absorber consists of a piston moving through a fluid such as oil. The upper part of the shock absorber is attached to the body of the car while the piston is connected to a spring. As the vehicle rides along a bumpy road, holes in the piston allow oil to move back and forth as the shock absorber vibrates up and down in response to the bump. Generally, the shock absorber is filled with a low-viscosity oil resulting in a slightly under-damped oscillation. Following the force of a bump, the body of the car will oscillate a few times about the equilibrium point before coming to rest.

Materials

Computer and appropriate software (e.g., Pasco's Data Studio)

Digital force probe

Graduated cylinder, plastic, 250- or 500-mL

Hooked mass, 200-g

Right-angle clamp

Ring stand, tall

Spring, 10 N/m

Thread, 1 m

Utility crossbar

Water

Safety Precautions

Although this demonstration does not pose any obvious risks, always follow laboratory safety rules. Clean up water spills immediately.

Preparation

Fill a 250-mL graduated cylinder nearly full with tap water. Use a right-angled clamp to attach a crossbar to a large ring stand. Secure a digital force probe to the cross bar and suspend a 10 N/m spring from the probe. Tie a 50-cm piece of thread to a 200-g hooked mass. Hang the mass directly from the spring and lower the mass alongside the water-filled graduated cylinder so that it rests next to the middle of the liquid.

Ask students what the position–time graph of a mass–spring system looks like. Activate the digital force probe and software. Pull the bottom of the spring until the mass touches the table and then release. Observe the graph in real time and discuss the consistency of the period of the system and how the amplitude appears to remain constant. Ask students what effect placing the mass into the water-filled cylinder will have on the graph. Activate the digital force probe and software. Pull the bottom of the spring until the mass touches the bottom of the cylinder and then release. Direct students' attention to the fact that, although the period decreases, it nevertheless is constant. Students should also notice that the amplitude decays very quickly, simulating the operation of a shock absorber.

Notes

Chapter 10

Fluids

Walking on Water

The surface tension of water is sufficient to support the weight of light objects.

Application

Surface tension • Unbalanced forces • Hydrogen bonding

Theory

The particles of water at the surface have special properties because they are subjected to unbalanced forces. Particles in the interior of the liquid are subject to attractive forces in all directions. However, particles at the surface of the liquid experience attractive forces only laterally and toward the interior of the liquid. As a consequence, particles at the surface experience a net force of attraction directed perpendicular to the surface towards the interior of the liquid. Water's surface tension is remarkably strong because it is a polar molecule that exhibits hydrogen bonding.

Hydrogen bonds are a type of dipole–dipole interaction that commonly occurs in polar molecules containing a hydrogen atom bonded to another highly electronegative element such as fluorine, nitrogen or oxygen. In such a molecule, the hydrogen atom is a center of positive charge while the nonmetal atom is a center of negative charge. Dipoles tend to align with their positive ends toward the negative ends of neighboring dipoles. This places the hydrogen atom in a bridge position between two nonmetal atoms, causing neighboring molecules to cluster together.

Small organisms such as water striders are able to float or "walk on water" since the downward pull of gravity is insufficient to break the attractive forces holding the water molecules at the surface together.

Hydrophobic substances, such as oil, that float on water, as well as emulsifiers, such as soaps and detergents, interfere with surface tension. Water striders are an indicator of water that has not been polluted with detergents.

Materials

Bar of soap

Battery jar or beaker, 2-L

Pepper

Plastic cherry tomato basket

Water

Safety Precautions

Although this demonstration does not pose any obvious risks, always follow laboratory safety rules. Clean up water spills immediately.

Demonstration

Floating Pepper

Fill a clean 2-L battery jar or beaker with tap water. Sprinkle pepper along the surface of the water. The pepper should float. Carefully dip the corner of the bar of soap into the surface of the water in the middle. The pepper will immediately shoot out toward the edges, away from the soap, and then settle toward the bottom of the container.

Sink the Ship

Fill a clean, 2-L battery jar or beaker with tap water. Carefully lower a cherry tomato basket onto the surface of the water. The basket should float. Carefully dip the corner of the bar of soap into the surface of the water. The basket should immediately sink like a rock.

Reference

Bilash, Borislaw; Shields, Martin; "Walking on Water"; *A Demo A Day—A Year of Biology Demonstrations.* Flinn Scientific: Batavia, IL, 2001; p 22. Reprinted with permission.

Battle of Two Syringes

Water is forced through a small syringe into a larger one demonstrating the hydraulic press effect.

<table>
<tr><td>Applications</td><td>Hydrostatics • Fluids • Pressure • Pascal's Principle</td></tr>
<tr><td>Theory</td><td>A hydraulic press consists of a small reservoir forcing fluid into a larger reservoir. The difference between the reservoir sizes results in a multiplying force or mechanical advantage due to the incompressibility of the fluid. With the pressure being equal throughout the system, the multiplying force is directly related to the ratio of the volumes of the two reservoirs. If two reservoirs of water are linked, then the pressure felt in one reservoir is just as great as the pressure in the other reservoir. If one reservoir's area is smaller than the other, then there is a multiplying effect on the force exerted on the larger reservoir by the smaller reservoir through the fluid. This effect is known as Pascal's Principle.</td></tr>
</table>

In this demonstration two different size syringes are connected using a small piece of tubing. When a force is applied to the smaller syringe, a larger force is generated in the larger syringe, that is, the smaller syringe is easier to push than the larger syringe. Consequently, the smaller syringe must be pushed a greater distance than the larger syringe moves.

Materials

Beaker for mixing water and dye and for filling syringes

Food coloring

Hot glue

Hot glue gun

Syringes, plastic, 2 different sizes (e.g., 5-cc and 50-cc)

Vinyl tubing, small piece, ¼″ id ½″ od, 3″ long

Water, tap

Safety Precautions

Although this demonstration does not pose any obvious risks, always follow laboratory safety rules. Clean up water spills immediately.

Preparation

Mix some food coloring with water in the beaker. Push the pistons of both syringes all the way down into the syringes. Glue the smaller of the two syringes to the tubing using the hot glue gun, but make sure not to impede the opening of the tube into the syringe. Fill the larger of the two syringes ¾ full with the colored water by drawing the fluid into the syringe. Connect the other end of the tubing to the larger syringe in the same manner as the first connection, making sure that no fluid escapes the syringe. Now, both syringes should be connected to the tubing and all the fluid is located in the larger of the two. Push the plunger of the larger syringe and allow the smaller syringe to fill about ¾ full with fluid.

Ask one of the larger students to take the larger of the two syringes in his hand, while finding a smaller student to take hold of the smaller syringe. Ask the smaller student to press the piston of the smaller syringe, while encouraging the larger student to keep the larger syringe's piston from moving out of the syringe. The larger student will find the task impossible. Ask the class to explain why. Discuss the use of hydraulic systems in everyday applications, including car brakes, heavy machinery, auto mechanic hoists, elevators, etc.

Optional: Mount syringes in a wooden frame with their plungers pointing upwards. Place a mass on the smaller syringe and another mass on the larger syringe so that the water and plungers are in equilibrium and not moving. It will be easy to illustrate how the pressure from the smaller mass on the smaller syringe is equal to the pressure from the larger mass on the larger syringe.

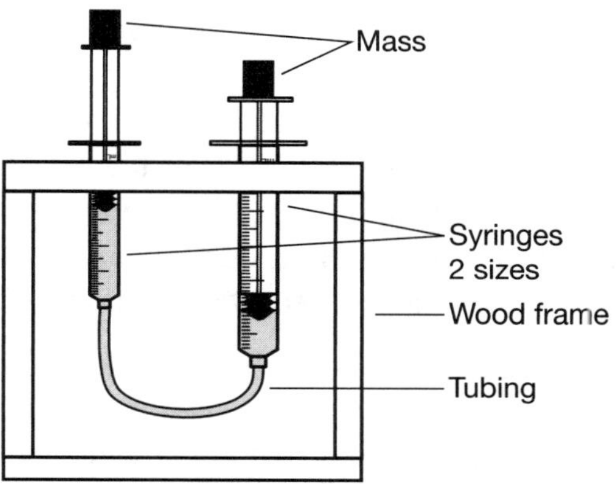

Atmosphere Keeps Me Down

A student is vacuum-packed in a large plastic bag, illustrating the large magnitude of Earth's atmospheric pressure.

Applications	Pressure • Atmospheric pressure • Force
Theory	The atmosphere of the Earth exerts a pressure of about 14.7 pounds of force per square inch or 101.3 kilopascal (kPa), where $1\ Pa = 1\ N/m^2$. This pressure is normally taken for granted since by and large the pressure is balanced by another pressure or force. For example, consider the effect of pressure on a marshmallow which consists of a sugary mixture laced with tiny air bubbles. When the marshmallow is placed in a vacuum jar and the air is removed from the jar, the marshmallow grows larger as the gas bubbles in the marshmallow expand.

In this demonstration a student's body is wrapped in a thick plastic bag. The bag is sealed around the neck of the student with the head protruding. A vacuum cleaner previously attached to the bag is used to extract some of the air from the bag and consequently lowers the air pressure in the bag. As the air pressure within the bag is lowered, the class will observe the bag tighten around the student due to the relatively higher pressure outside the bag. The difference in pressure will be so dramatic that it will prevent the student from moving her arms outward. In fact, the student will be vacuum-packed like a bag of coffee.

On average, the human body has a surface area of about $1\ m^2$. Consequently, the quantity of air pushing on a body amounts to about 1000 kg! The reason this pressure is not noticed is due to the counterbalancing forces of the body including turgid pressure of the cells and blood pressure.

Materials	Garbage bags, construction, plastic, 3 mm thick
	Rubber band, $4'' \times \frac{1}{4}''$
	Tape, duct
	Utility knife
	Vacuum cleaner with long hose
Safety Precautions	Since the student volunteer will be placed in a confined environment, verify that the student volunteer does not have a history of claustrophobia. Discontinue the demonstraton immediately if the student expresses discomfort or anxiety. Always follow all laboratory safety rules.
Preparation	Use a utility knife to cut a small hole the size of the vacuum cleaner hose in a garbage bag. Insert about $6''$ of the end of the hose through this hole into the bag. Use duct tape to seal the bag around the hose.

Invite a student of small stature to assist with the demonstration. Have the student sit into the large plastic construction bag with the hose facing the audience. Ask the student to sit inside the bag with his legs crossed and holding the end of the hose with both hands to prevent its obstruction during the demonstration. Bring the opening of the bag against the neck of the student and carefully seal it around the neck using a $4'' \times \frac{1}{4}''$ rubber band. The seal should be snug, but the student should still be comfortable. Turn the vacuum on and start removing air from the bag. The bag will start to press against the individual's body and feel tighter against him. Within a minute, the student should have a difficult time moving both arms and legs. The student may be released by disconnecting the vacuum. Discuss the amount of air constantly pressing on our bodies and why we do not feel it.

References

Special thanks to Harry Rheam, Eastern Senior High School, Atco, NJ for sharing this idea.

Fingering Out Buoyancy

The buoyant force acting on a finger dipped in water is measured using a balance.

Applications Buoyancy • Density • Newton's Third Law

Theory The concept of a buoyant force is often illustrated with an object floating on a fluid. For example, a ship floating on water exerts a downward force on the water as much as the water is providing a force on the ship. The upward force of the water, called the buoyant force, is equal and opposite to the weight of the ship. The difference between a ship that would float on water and a ship that would sink depends on the ratio of the densities between the water and the ship itself. A ship which is denser than the water will sink, whereas a ship which is less dense will float. As cargo is added to a floating ship, its density increases. As its density increases, the ship becomes suspended lower into the water and the water line on the hull will move up toward the deck of the ship. In terms of force, as the weight of the ship increases, the downward force on the water increases. Consequently, the upward force—the buoyant force—must increase to keep the ship afloat. In this demonstration a finger is dipped into a beaker of water resting on a balance. The buoyant force registers on the balance due to the action–reaction force pairs as the finger is pushed into the water. In this case, the buoyant force is the action, whereas the reaction is the downward push of the finger into the water as shown by the increase on the balance.

Materials Balance, centigram, mechanical or digital

Beaker, 600-mL

Mass set for mechanical balance, approximately 600 g

Water

Safety Precautions Although this demonstration does not pose any obvious risks, always follow laboratory safety rules. Clean up water spills immediately.

Demonstration Fill the beaker ¾ full with water and place on a balance. If using a mechanical balance, add masses to the other side of the balance until it is in equilibrium. Ask students to predict what the balance scale will read if you insert a finger into the water without touching any other part of the beaker or balance. Proceed by carefully placing most of your finger into the water. Invite a student to serve as the "class eyes" and ask him or her to report the reading on the balance before and after the finger is inserted into the water. Show students that if you insert your finger further into the water, the reading on the scale increases. Review the free-body diagram of the force of the finger acting on the water and the buoyant force acting on the finger.

References

Neie, Van E. "Beware of Greeks Bearing Pan Balances"; *The Physics Teacher;* 1995; vol 33, p 6.

Ward, David A. "Finding the Buoyant Force"; *The Physics Teacher;* 1994; vol 32, pp 114–115.

Ward, David A. "Letter: Ward Responds"; *The Physics Teacher;* 1995; vol 33, pp 6–7.

Giving Soda Pop a Lift — A Twist on an Old Idea

A can of regular soda pop sinks in water while diet soda pop floats, but they both float in salt water.

Application	Buoyancy • Density
Theory	Why do objects float or sink in a medium? It depends upon the relative density between the object and the medium (generally, a fluid) in which it is immersed. The displacement of the surrounding medium by the object plays a key role in determining if the object sinks, floats or has a neutral buoyancy. If the mass of the displaced medium is greater than the mass of the object, then the object sinks because its weight is greater than the buoyant force provided by the medium. In contrast, an object will float if its mass is less than the mass of the displaced medium. An object is said to have neutral buoyancy when the mass of the medium displaced is exactly equal to the mass of the object in the medium. In this case, the object can be placed anywhere in the fluid and not have any net force exerted on it, and will neither float nor sink. In this demonstration a can of regular soda pop and a can of diet soda pop are placed into water. The regular soda pop sinks, while the diet soda pop floats due to the different sweeteners that they contain. Regular soda pop contains about 40 g of corn syrup per serving, while diet soda contains about 0.2 g of Aspartame—a potent artificial sweetener.
Materials	Aquarium, 5-gal or similar size container (e.g., 2 gal empty plastic pretzel container) Pop, carbonated soda, diet, 12-oz can Pop, carbonated soda, non-diet, 12-oz can Table salt, 26 oz Water *Note:* Not all 12 oz. cans of diet soda pop float, and not all regular soda pop sinks. The drinks that you choose should be tested before doing the demonstration in front of the class.
Safety Precautions	Although this demonstration does not pose any obvious risks, always follow laboratory safety rules. Clean up water spills immediately.
Demonstration	Hide the container of salt from the class and fill the aquarium 9/10 full of water. Have the class vote on which soda will sink or which will float in the water before immersing the cans into the water. Ask the students to explain their theories. Begin by placing the regular soda pop in the water first. Place the can sideways in the water to allow any air captured by the "dimple" in the bottom of the can to escape, as this can impede the demonstration. The can of regular soda pop should sink in the water.

Have the class explain the observation in terms of the densities of the soda pop and water. Now place the diet soda can in the water. This soda should float. Again, ask students to explain the behavior.

After discussing the densities of the pop and the water, ask the students how they would bring the regular soda pop to the top of the water without physically "grabbing" the soda and bringing it to the surface by hand. After coming up with some ideas, introduce the idea of changing the density of the fluid by adding salt to it. Now take the container of salt and start pouring it into the container, being careful not to pour it directly onto the soda under the surface. As the salt starts to mix with the water, the soda pop at the bottom will begin to stir and slowly rise to the surface. There is even a point, if you are careful, where neutral buoyancy can be exhibited and the soda will float in the middle of the salt–water solution. As more salt is added and the salt mixes with the water, the can of regular soda will finally rise to the surface of the solution next to the diet soda, although it will not float as high in the solution as the diet soda.

Ask students if they would feel more of a buoyant force in the ocean or in a clear swimming pool and then ask them to explain why. Ask the students who have been in both to discuss the sensation of one type of fluid's buoyant force versus the other.

Reference

Mellen, Walter Roy. "Oscillations of Eggs and Things: Behavior of objects in fluids having densities that increase with depth"; *The Physics Teacher;* 1994; vol 32, pp 474–475.

Balls in Beans

A mixture of three materials is made "fluid," causing them to separate according to their densities.

Application	Density • Buoyancy

Theory

The density of an object is the ratio between the mass and volume of the object. An object placed in a medium such as water will either float, sink or hover in the fluid depending on its relative density compared to water. In this demonstration a mixture consisting of beans and two balls of identical volumes but different masses—a Ping-Pong ball and a steel ball—will serve as a model of this principle. This mixture of solids is agitated to simulate a fluid mixture. As the mixture becomes fluid, it will rearrange itself according to the relative densities of the three components. This model is akin to observing a helium balloon and a balloon filled with carbon dioxide in air. The helium-filled balloon will float while the carbon dioxide-filled balloon will sink. Similarly, a cork floats on water while a rock will sink.

In the case of a mixture of objects of equal densities, the objects will layer themselves according to their relative volumes with the larger pieces towards the top of the mixture. This is commonly observed in packages of food including potato chips, breakfast cereal, and popped popcorn—the biggest pieces are usually found at the top.

Materials

Ping-Pong balls, 2

Pinto beans, uncooked, 1-lb bag or similar, 5–10 bags

Pretzel container, plastic, 2-gal or transparent or semitransparent plastic salad or punch bowl

Steel ball, equal in diameter (40 mm) to Ping-Pong ball

Tape, transparent

Utility knife

Note: A brass or lead ball may be used, however, it should be painted to contrast with the red pinto beans.

Safety Precautions

The materials in this demonstration are considered safe. Follow all laboratory safety rules.

Preparation

The materials in this demonstration are considered safe. Follow all laboratory safety rules. Use a utility knife to cut a small opening in one Ping-Pong ball. Carefully push beans through this opening until the ball is filled with beans. Cover the hole with tape.

Pour 5–10 bags of beans into an empty plastic pretzel container or transparent plastic salad bowl. Place the uncut Ping-Pong ball about 5 cm under the surface of the beans and place one metal ball on the surface.

<table>
<tr><td>Demonstration</td><td>Direct students' attention to the container of beans with the metal ball resting on top. Ask students to predict what will happen if the container is vigorously agitated while in contact with the table. Students will observe that as the container is shaken, the metal ball will sink into the pool of beans while the hidden Ping-Pong ball rises above the surface. Expect students to be astonished with this sight. Capitalize on their curiosity by having them explain how the demonstration works in terms of density.</td></tr>
</table>

Demonstration

Direct students' attention to the container of beans with the metal ball resting on top. Ask students to predict what will happen if the container is vigorously agitated while in contact with the table. Students will observe that as the container is shaken, the metal ball will sink into the pool of beans while the hidden Ping-Pong ball rises above the surface. Expect students to be astonished with this sight. Capitalize on their curiosity by having them explain how the demonstration works in terms of density.

Remove the balls from the container and pass them around the class along with the bean-filled ball. Students should notice that although the Ping-Pong ball and steel ball have equal volumes, their masses differ significantly. In other words, their densities are different.

Place the bean-filled ball about 5 cm under the surface of the beans in the container. Ask students to predict what will happen to this ball as the container is shaken. Students will see that this ball will rise towards the surface. This behavior is analogous to the larger chips or cereal bits rising to the top of a bag or box and is a result of how similar density materials behave when they are different sizes.

References

Prigo, Robert B. "Liquid Beans"; *The Physics Teacher;* 1988; vol 26, p 101.

Winter, Rolf G. "On the Difference between Fluids and Dried Beans"; *The Physics Teacher;* 1990; vol 28, p 104.

Streaming Toward a Candle

A candle is extinguished in a novel way illustrating the Coandă effect.

Application	Coandă effect • Streamline flow • Turbulent flow • Fluid dynamics
Theory	In watching the movement of water in a slow-moving stream, it's fairly easy to observe how a fluid approaches and then moves around the obstacles in its path. Large rocks, protruding branches, and bends in the streambed affect the flow of the water. The smooth movement of this flow as it passes a small rock is known as laminar or streamline flow.

When a fluid such as water travels down a curved channel, the water exerts a force on the sides of the curve while the curved surface exerts a normal force onto the water, thereby bending the streamline. At low velocities, streamlines do not intersect one another. When a streamline encounters an obstacle, such as a rock, the streamlines diverge, follow the surface of the obstacle, and then recombine once past the obstacle. This is known as the Coandă effect. The Coandă effect describes how a fluid will tend to follow a curved surface of a solid due to the surface tension or Van der Waals forces of attraction between the fluid and solid. In contrast, a fluid exhibits turbulent flow when the fluid loses its streamline characteristics, due to either an increase in the speed of the flow or a dramatic change in the surface or size of the obstacle it encounters. A rock protruding from a fast-moving stream will exhibit turbulent eddy currents on its downstream side, evidenced by a frothy mix of water. Turbulent flow is a chaotic behavior, which is not well understood in the study of fluid dynamics.

Scientists use these ideas to explore the fluid dynamics of many different systems. Automobile engineers regularly put their models of cars under development in wind tunnels where they introduce lines of smoke, known as "streamers," to illustrate the streamline flow around and over their car models. The goal is to avoid turbulent flow, which is an indication of increased friction or drag on the car. Drag results in decreased fuel efficiency.

Materials	Candle with candleholder, $\approx$ 15 cm tall
	Lighter
	Soda bottle, plastic, 2-L, with label removed; filled with water and capped
Safety Precautions	Use caution when working with flames. Always follow all laboratory safety rules.
Demonstration	Place a candle on a table about 50 cm from the edge of the table. Light the candle and ask the students what will happen if you blow towards the candle. Of course, most will think you can blow out the candle. Confirm their prediction by blowing out the candle. Discuss the movement of the air toward the candle.

Now place the soda bottle midway between you and the candle. Again, ask students if you can still blow out the candle. This time there may be some disagreement about whether you still can extinguish the candle. Relight the candle and, once again, placing your head level with the candle but on the opposite side of the bottle from the candle, proceed to blow out the candle. This may require some practice. Ask students to describe the flow of fluid (air) from you to the candle.

Curve Baller

A Styrofoam® ball is launched from a homemade jai alai paddle to illustrate how balls in sports can be made to curve.

Application Coandă effect • Bernoulli effect • Magnus effect • Fluid dynamics

Theory In 1672, Isaac Newton noted how a spinning tennis ball curved when moving forward in the air. Newton proposed that the curved path was somehow associated with the spin of the ball. Later, others investigated this phenomenon in great detail. Our most recent understanding of how a ball curves as it moves through air is explained by a combination of the Coandă, Bernoulli and Magnus effects. The Coandă effect describes how a fluid will tend to follow a curved surface of a solid due to the surface tension or Van der Waals forces of attraction between the fluid and solid. In the case of a baseball (or golf ball, airfoil, etc.), the layer of air closest to the ball—the boundary layer—slips around the ball in concert without breaking up (see Figure 1). According to the Bernoulli effect, if the ball is spinning (see Figure 2) as it travels through the air, a series of vortices trail behind the ball and are slightly pulled along the surface of the ball in the direction of the spin. This causes an imbalance of pressure, with a high pressure located where the spin meets the opposing streams of air—where the vortices collect (A)—and a relative low pressure on the opposite side of the ball (B) where the spin and air streams are unidirectional. The resulting pressure differential provides a force known as the Magnus effect that pushes the ball from the high to low pressure region.

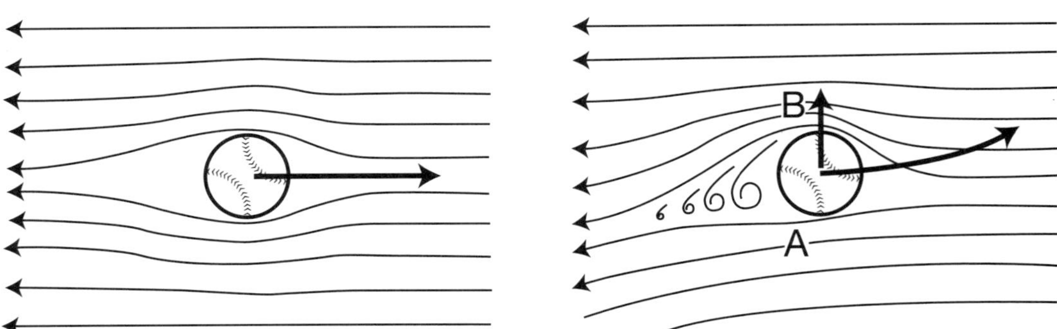

Figure 1. Ball not spinning   **Figure 2.** Spinning ball

The explanation of how a ball is made to curve is often limited to the Bernoulli effect; however, a complete explanation requires the inclusion of the Coandă and Magnus effects.

This demonstration uses a home-made paddle modeled after a jai alai paddle to launch a Styrofoam ball. Jai alai is a fast paced game in which a ball is launched and caught in a scoop-like basket. As the ball is launched, the basket causes the ball to rotate and curve as it moves through air.

Adhesive, construction

Brad nails, 1″, 10

Hand saw

Masonite boards, 3/16″ × 3″ × 24″, 2

Sandpaper, medium-grit sheet, 8.5″ × 11″

Spray paint, Styrofoam®-safe, any bright color, 1 can

Styrofoam ball, 2″

White glue

Safety Precautions

Never throw balls at the class. Throw the ball away from or parallel to the class to minimize the possibility of hitting someone in the eye. Always follow manufacturers' directions when using power tools. Always follow laboratory safety rules.

Preparation

Use a hand saw to cut out a 2″ × 6″ section from each board as illustrated in the diagram. Use construction adhesive to make a lap joint between the 24″ lengths of the boards, forming a right angle. Use 8–10 brad nails to secure the board. Apply a bead of construction adhesive along the entire length of the inner seam of the joint. Use white glue to attach sheets of sandpaper to the entire length of the inside fold of the paddle. Finally, cut 4–5 small triangular pieces (right isosceles triangles) and glue these inside the handle of the paddle to reinforce it. Allow the adhesive to cure according to the manufacturer's directions. Use Styrofoam-safe spray paint to color the Styrofoam balls for better visibility.

Demonstration

Place the ball in the paddle near the handle and with one quick smooth motion fling the ball outward, like a tennis racket hitting a ball. Expect the ball to move forward curving to the right if swinging with your right hand. The ball will curve left if using your left hand. With some practice, this effect and the accuracy of your "curve balling" can be increased. You may also want to attempt different angles of throws, and try to get the Magnus effect to impart an upward, left or downward force. Just practice a bit and all these curvatures can be demonstrated.

References

Barnes, George "Demonstrating Curved Trajectories of a Spinning Ball"; *The Physics Teacher;* 1981; vol 19, p 403.

Briggs, Lyman J. "Effect of Spin and Speed on the Lateral Deflection (Curve) of a Baseball"; "Magnus Effect for Smooth Spheres"; *American Journal of Physics;* 1959; vol 27, pp 589–596.

Smith, Norman F. "Bernoulli and Newton in Fluid Mechanics"; *The Physics Teacher;* 1972; vol 10, pp 451–455.

Float Your Ball!

A stream of air is directed upward at a beach ball suspending it in the air.

Application | Coandă effect • Magnus effect • Fluid dynamics • Lift

Theory | In this demonstration a beach ball is placed in the stream of air emitted by a leaf blower directed upwards. The behavior of the ball is due to a number of effects and combination of forces. First, the ball rises upward due to the force of air pushing on the bottom of the ball. The ball will hover at a distance above the blower once the upward force is balanced by gravity. The stability of the ball within the stream may be explained by the Coandă effect: As a fluid flows past a curved surface, there is a drag effect on the fluid due to Van der Waals force between the fluid and surface. In the case of the beach ball, a column of air emitted by the blower flows evenly around the ball and then reunites if the ball is in the middle of the air stream. If the ball is pushed away from the center of the column, the flow of air increases on one side of the ball, resulting in a net force back toward the center of the column. The net force is known as the Magnus effect.

When the column of air is tilted—as much as 45°—the ball will still remain suspended within the column. Tilting the column is akin to pushing the ball to the side of a vertically-aligned column of air as described above. However, when tilted, gravity pulls the ball downward slightly, bringing the ball closer to one side of the column (nearer the ground) and producing a greater air flow along the top surface of the ball resulting in a net force back towards the center of the column.

Materials | Beach ball, 12″–24″ in diameter

Leaf blower, battery operated is ideal

Safety Precautions | Although this demonstration does not pose any obvious risks, always follow laboratory safety rules.

Demonstration | First, hold the beach ball in front of the students, and then ask them to explain the forces acting on the ball when it is let go. Next, turn on the leaf blower and direct the flow upward. Using your free hand, place the ball in the air stream. This may require some practice, as the location for placing the ball into the air stream will vary according to the type and size of the beach ball and the output of the blower. The ball will usually oscillate in the stream at first, but will quickly settle into a stable spot.

Ask students to explain the forces acting on the ball as it sits in the stream of air. If you push on the side of the ball lightly, you'll note that as the ball approaches one side of the stream, it will quickly be pushed back into the center of the stream of air, once again becoming stable. Next, slowly redirect the flow of air toward the horizontal (as far as 45°) and the ball should follow along the flow, staying suspended. Again, ask your students to explain the forces on the ball.

<table>
<tr><td>References</td><td>

McDonald, Kirk T. "Levitating Beach Balls"; *American Journal of Physics;* 2000; vol 68, pp 388–389.

Raskin, Jef. "Model Airplanes, The Bernoulli Effect Equation, and the Coandă Effect"; http://jef.raskincenter.org/published/coanda_effect.html; Accessed November, 2008.

Weltner, Klaus; Ingelman-Sundberg, Martin. "Misinterpretations of Bernoulli's Law"; University Frankfurt, Frankfurt: Germany; http://user.uni-frankfurt.de/~weltner/Mis6/mis6.html; Accessed November, 2008.

</td></tr>
</table>

Smoke Ring Cannon

Smoke rings are launched across the room using a homemade device to illustrate a unique motion of fluid.

Application

Vortex movement • Fluid dynamics • Streamline flow • Turbulent flow

Theory

The spinning movement of water flowing down a drain is called a vortex. Other examples include the movement of air off an airplane wing or in a tornado as it rotates, and the movement of air into the eye of a hurricane. Many vortices can be the result of turbulent flow—the chaotic movement of a fluid, though streamline flow can also exhibit vortex characteristics.

In this demonstration vortices of smoke are emitted from a homemade cannon. Smoke is pushed from the rear of the cannon and moves in a streamline fashion towards the opening of the cannon. The streamlines follow the front inner-surface of the cannon and rush toward the center opening. At the center the streamline flow transforms into a turbulent flow, resulting in the formation of a vortex. The vortex takes the shape of a toroid—a donut shape—with the smoke swirling into the hole of the toroid. The swirling movement of smoke is similar to the air currents moving in the eye of a hurricane.

Materials

Bucket, plastic, 5-gal

Cabinet handle with 2 screws

Candle on stand

Drill

Drill bit, ¼″

Fogger with fogger fluid, available from party stores, some electronic catalogs, and some department stores

Lighter

Spandex® fabric or latex rubber, 2′ × 2′ section

Tape, duct

Utility knife

Safety Precautions

Although this demonstration does not pose any obvious risks, always follow laboratory safety rules. Follow manufacturers' directions whenever using power tools.

Drill two holes into the center side of a 5-gallon bucket and then fasten a kitchen cabinet handle to the bucket. Use a pencil and compass to draw a 10-cm circle in the center of the bottom of the 5-gallon bucket. Use a utility knife to cut through the plastic of the bucket. Use the knife to remove any remaining burrs so that the hole is as smooth as possible. Stretch the fabric or rubber tight across the top opening of the bucket so that the fabric is under some tension, and fasten to the bucket using duct tape.

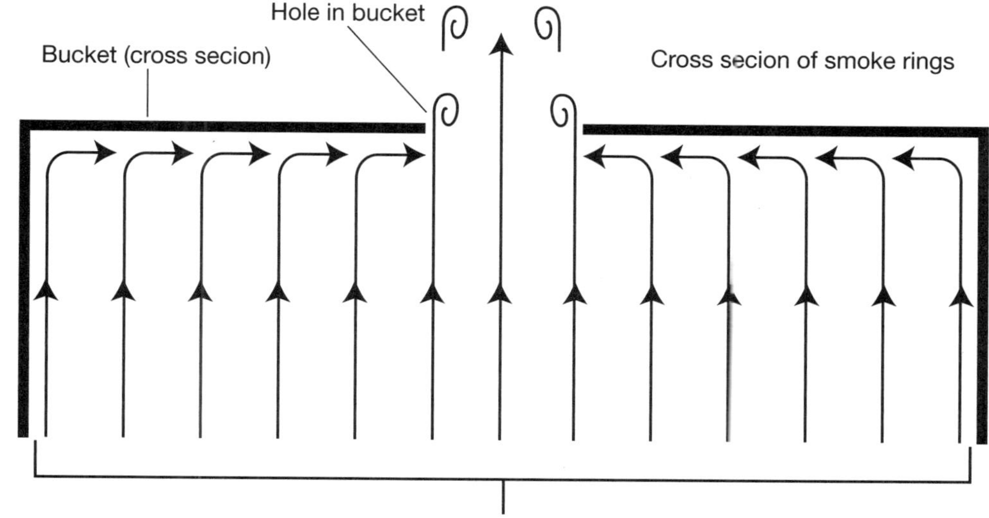

Demonstration

Prior to class, turn on the fogger and let it warm up. Begin the demonstration by lighting a candle situated about 3 meters away from the cannon. Hold the cannon so that the small hole is aimed at the candle, then slap the rubber or fabric end of the cannon with one hand to force air down the cannon. If your aim is good, the "projectile" of air will blow out the candle. This is something you should practice before attempting in front of your audience. You could also direct the cannon at individuals in your class—perhaps using the blasts from the cannon to flick students' hair or scatter their papers.

Ask students to explain the movement of the air prior to and after moving through the opening of the cannon. Discuss how the streamline flow transforms to a turbulent flow, resulting in the formation of a vortex.

Now, explain that in fluid dynamics a marker such as smoke or ink is used to visualize the streamline and/or turbulent flow of the fluid. Proceed to fill the cannon through its opening with some smoke from a fog machine and repeat the demonstration. Direct students' attention to the toroid shape of the air projectile and to its backward spin as it moves across the room. Notice also that the toroid also rotates around its center.

Notes

Chapter 11

Waves and Sound

Types of Waves

A Slinky® is used to illustrate the properties of a transverse wave versus a longitudinal wave.

Application | Transverse waves • Longitudinal waves • Energy transfer • Amplitude

Theory | There are two types of waves—transverse and longitudinal waves. Waves are disturbances that travel through space or a medium, transferring energy from one place to another. In the case of a transverse wave, a disturbance, such as a vibration, occurs at a right angle to the propagation of the wave. A vibrating string, light waves, radio waves, and heat waves are examples of transverse waves. In contrast, the disturbance in a longitudinal or compression wave is in the same direction as the propagation of the wave. Sound and the primary waves of earthquakes (also called P-waves) are examples of longitudinal waves. Earthquakes also produce a transverse wave known as the S-wave. During an earthquake, the S-wave travels faster and is generally a more gentle wave than the following P-wave. The degree of movement caused by the S-wave may be used as a warning since it forecasts the intensity of the approaching P-wave. In those waves that move through a medium, such as sound waves, heat waves or water waves, the individual particles that make up the medium do not themselves travel with the energy. The particles instead oscillate from side to side transferring the energy from one particle to the next—akin to a line of dominos falling over. However, the particles spring back to their original position. A water wave exhibits both transverse and longitudinal properties. As the wave travels through the water, the water particles travel in clockwise circles—exhibiting both vertical (transverse) and horizontal (longitudinal) movement.

Materials | Plastic cup, disposable, 5–16 oz (e.g., Dixie® or Solo® brand)

Triple-length Slinky®

Safety Precautions | Although this demonstration does not pose any obvious risks, always follow laboratory safety rules.

Demonstrations | **What Is a Transverse Wave?**

Stretch a triple-length Slinky along the smooth surface of a floor or on top of a long table. The Slinky should be stretched a length of approximately 5 meters with one person holding one end and another person holding the other end.

Have one person generate a transverse wave pulse by applying a quick snap of the wrist done at a right angle to the stretched Slinky. The other person should simply hold the other end of the Slinky steady. Produce a number of waves, one at a time, to allow students to see how the pulse behaves as it moves away and then back toward you. Ask students to summarize what they observe as the pulse moves. Point out that the wave moves or propagates at a right angle to the generation of the pulse. Also note that the wave flips sides as it reflects off the hand holding the opposite end of the Slinky.

Finally, discuss what sort of energy or phenomena move in the form of transverse waves.

Waves Transport Energy

Discuss the idea that the Slinky serves as a medium that transfers energy from one point to another. Demonstrate this by using a pulse to move a plastic cup. Begin by placing a plastic cup on the floor next to the stretched out Slinky. The cup should be approximately 30–40 cm away from the Slinky. A single transverse pulse of the slinky will move the cup, showing that a disturbance may be created at one end of the Slinky (a medium) and causes an object positioned elsewhere to move. Notice how the cup always moves in a path perpendicular to the propagation of the wave. Point out that although the energy was transferred to the cup, the coils of the Slinky return to their original positions.

What Is a Longitudinal Wave?

As with the illustration of a transverse wave, stretch a triple-length Slinky along the surface of a smooth floor or on top of a long table.

While holding one end of the Slinky, bring together 10–20 coils of the stretched Slinky and then release the bunch. A compression wave will move through the Slinky from one end to the other. Produce a number of waves, one at a time, by pushing and pulling the Slinky rapidly to allow students to see how the pulse behaves as it moves away and then back towards you. Ask students to summarize what they observe as the pulse moves. Point out that a longitudinal wave moves in the same direction as the propagation of the wave.

Finally, discuss what sort of energy or phenomena move in the form of longitudinal waves.

Anatomy of a Wave

Generate a series of transverse pulses and point out the crest and trough of the wave. Create waves with small wavelengths, followed by larger wavelengths, while keeping the amplitude constant. Ask students how they would measure the difference between the two sets of waves. Students should understand that a wavelength is the measurement of one complete wave cycle—typically measured from crest to crest or trough to trough.

Generate another series of transverse pulses—first, a series of small waves followed by ones with a larger amplitude, while keeping the wavelengths constant. Ask students if they recognize the difference between each series, and if so, have them describe the difference to the class. Be sure that students recognize that the amplitude of the wave is directly related to the disturbance created by the motion of your hand. The amplitude of a longitudinal wave may also be shown, although the demonstration may be less dramatic and therefore not as effective.

Play That Funky Sound

The medium through which sound travels does not travel with the sound.

Application | Longitudinal waves • Sound media

Theory | The disturbance in a longitudinal or compression wave is in the same direction as the propagation of the wave. However, the media through which the wave travels does not travel with the wave. As a longitudinal wave travels, it compresses the medium along its path. Once the medium is compressed, it usually snaps back or rebounds towards its original position. If this were not true, then a person speaking would cause quite a wind storm in the direction of their voices. In this demonstration a bubble film covers the bell of a trumpet. Although sound passes through the bubble film, the air in the trumpet does not.

Materials | Soap bubble solution

Shallow pan (e.g., pie plate)

Trumpet, coronet or bugle (brass or toy version)

Safety Precautions | Wear chemical splash goggles when performing this demonstration. Always follow laboratory safety rules.

Demonstration | Pour some soap bubble solution into a shallow pan. Dip the bell of the trumpet into the solution forming a bubble film over its end. Play a short tune. Ask students how the sound left the trumpet and got to the back of the room.

Reference | Hults, Morris G. "Sound Waves"; *A Potpourri of Physics Teaching Ideas;* American Association of Physics Teachers: College Park, MD; p 248.

A ringing bell is placed into a container filled with air, without air, and then filled with other gases.

Application

Speed of sound • Density of gases • Frequency

Theory

Sound waves require a medium through which to travel. Sound waves travel in air at a rate of about 344 m/s at 20° C. The speed of sound is dependent upon many factors including temperature and the nature of the media. If no media is present, then sound cannot travel. Sound cannot travel in the vacuum of outer space. The speed of sound in a gas is proportional to the square root of the gas's density (ρ). Consequently, in helium ($\rho = 0.1786$ g/L) sound travels 2.7 times the speed of sound in air, while in sulfur hexafluoride ($\rho = 6.13$ g/L) sound travels at 0.44 times the speed of sound in air. Sound travels faster in liquids and fastest in solids. The difference in speeds is due to the elasticity of the media. The greater the elasticity, the faster the compressed molecules can snap back into place.

When one speaks with lungs filled with helium gas, the result is an amusing high pitched voice reminiscent of a cartoon character. Although the frequency of the vocal cords is not altered, the sound produced by the cords moves more quickly through the helium-filled vocal tract, giving the sound leaving the mouth a high pitch or frequency. Some mixtures of breathing gas used in deep sea diving contain up to 79% helium to reduce the risk of nitrogen narcosis and oxygen toxicity when breathing at these depths. Breathing such a mixture will produce the same effect— shifting your voice to a higher pitch. In contrast, speaking with lungs filled with the relatively inert gas, sulfur hexafluoride will be equally amusing. The high density of this gas will cause the pitch of one's voice to be considerably lower.

Materials

Helium gas, 2-L (small disposable helium tanks may be purchased at party stores)

Sulfur hexafluoride gas, 2-L

Balloons, latex, 2

Battery, 6-V

Bell jar sound apparatus*

Connector wires with alligator clips, 2

Vacuum pump, electric or hand-powered

Note: Some science suppliers sell bell jars equipped with buzzers instead of a traditional electric bell. Using a "buzzer jar" rather than a bell jar misses out on the opportunity to see the bell ring but not hearing it when the jar is evacuated—an important feature of this classic demonstration. If your bell jar contains a buzzer, consider replacing the buzzer with an electric bell.

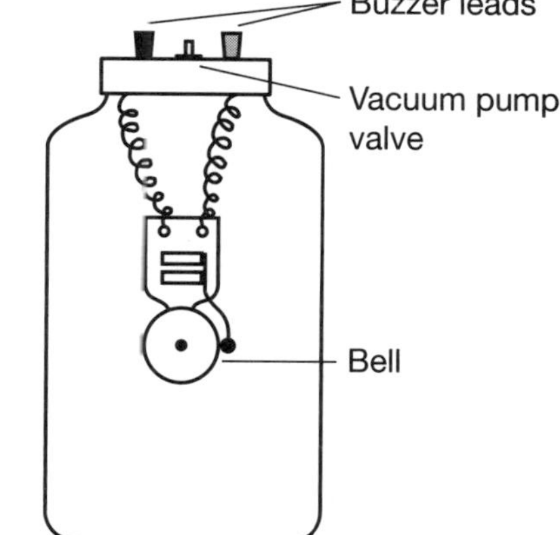

Both helium and sulfur hexafluoride are inert gases. When taken into the lungs, however, the gases prevent oxygen from being absorbed by the body, which can lead to hypoxia and asphyxiation. As a general rule, lungs should not contain either of these gases for a period longer than half the time that you can hold your breath. The demonstration also should not be immediately repeated. Prior to each attempt, it is advisable to take a few deep breaths. Do not attempt this demonstration with any other gases. Do not permit students to breathe in the gases. Helium and sulfur hexafluoride are among the few gases that can be safely used in such a demonstration. Other gases may be absorbed into the body and cause illness or even death.

Preparation

Inspect the bell jar apparatus and verify that the bell and its housing are suspended by wires so it is not in contact with the jar itself.

Demonstrations

Sound Production in Air versus a Vacuum

Connect the bell of the bell jar apparatus to a 6-volt battery and direct students' attention to the fact that the hammer is striking the bell and produces sound. The electric arc at the connection between the hammer and the electromagnet may also be visible. Ask students what will happen if the air in the jar is removed. After a short discussion, proceed to evacuate the jar using an electric or hand-powered vacuum pump. As the air is removed, students will notice that the volume of the bell will diminish. Stop the vacuum pump and direct students' attention to the fact that the hammer of the bell continues to strike the bell, yet little or no sound is evident. Ask students what will happen if the air is returned to the container. Following discussion, slowly return the air to the jar. The increase in volume should be apparent and dramatic with the vacuum pump off.

The Effect of Density on Sound Production

Ask students what will happen if the bell is placed in a different gas such as helium. Following discussion, remove the bell from the bell jar apparatus. Ask a volunteer to hold the empty jar so that the opening is facing downward. Fill the jar with helium gas from a helium tank. Connect the bell using connector wires with alligator clips to a 6-volt battery and move the bell into the helium-filled jar. The sound emitted from the apparatus will have a higher frequency when the apparatus is immersed in the helium atmosphere.

Repeat the demonstration using sulfur hexafluoride gas, but this time fill the jar with the opening facing upward. The sound emitted from the apparatus filled with the denser gas will have a frequency considerably lower than that heard when it was immersed in air or helium.

Funny Voices

Fill one balloon with helium gas. Take a few deep breaths of air first and then draw the helium from the balloon into your lungs. Immediately begin explaining what is going on using your helium-laced voice. Do not hold your breath with helium-filled lungs. After a minute of talking, take a few deep breaths of air again to ensure that the helium has been flushed out of your lungs.

After a few minutes, repeat the demonstration using sulfur hexafluoride gas. Once again, after a minute of talking, take a few deep breaths of air to ensure that the gas has been flushed out of your lungs.

Reference

The voice demonstration is based on a performance by Professor Julien Sprott performed at the Institute of Chemical Education at the University of Wisconsin–Madison in the summer of 1989.

Standing Wave Generator

An electric hand drill is used to generate a standing wave to illustrate transverse waves and the relationship between harmonics and frequency.

Applications Standing wave • Resonance • Transverse waves • Harmonics • Frequency

Theory In this demonstration a standing wave is created in a string using a hand drill. A string is connected to a modified chuck. As the chuck rotates, the drill imparts a series of transverse wave pulses. When the initial pulse leaves the drill, it travels down the string and then, like an echo, reflects off of the fixed end (the ring stand) back toward the drill. The initial pulse continues to move back and forth along the string while new wave pulses are generated by the drill. If the frequency of the pulses is matched to the velocity of the pulses, a standing wave becomes visible. It is called a standing wave because the wave does not appear to be traveling. The simplest standing wave consists of a single loop. The single loop represents both the crest and the trough rapidly exchanging places. This is also called the first harmonic or the natural frequency of the string. The makeup of the string (length, density, diameter) and the tension all contribute to its natural frequency. If the speed of the drill is slowly increased, two loops become visible, corresponding to the second harmonic of the string system (i.e., one crest and one trough). The two loops are separated by a node. The node corresponds to the point in the string where the pulses moving to and from the drill cancel each other out in a process called destructive interference. Increasing the speed will produce higher harmonics. While the string is vibrating, touch the node of the string to show that the energy still travels through the rope since touching the node does not alter the original waveform. The motion of the string can be "frozen" using a strobe light tuned to the frequency of the string. Natural frequencies are also called resonant frequencies.

Materials

C-clamps, large or adjustable, 4	Propane torch
Electric hand drill	Rheostat transformer (e.g., Variac®)
Eyebolt, 5″ × ¼″	Ring stand, small, with ring
Mass, 1 or 2 kg, with hook	Rope, ¼″, fluorescent-colored, 2 m
Pipe clamps, 3″ id	Vise
Pliers, large or pipe wrench	Wooden board, 2″ × 4″ × 12″
Plywood, 12″ × 12″ × ¾″	Wood screws, 3″, 3

Optional: Adjustable strobe light (with tachometer, if available)

Safety Precautions Follow manufacturers' recommendations when using power tools. Propane gas is highly flammable—keep cylinders away from heat and protect from falling. Wear safety glasses when performing this demonstration and follow all laboratory safety rules.

Preparation

Secure an eyebolt in a vise. Use a propane torch to heat the middle of the stem of the bolt for about 1 minute. Use a large pair of pliers or a pipe wrench to bend the bolt into a right angle at its center. Reheat the

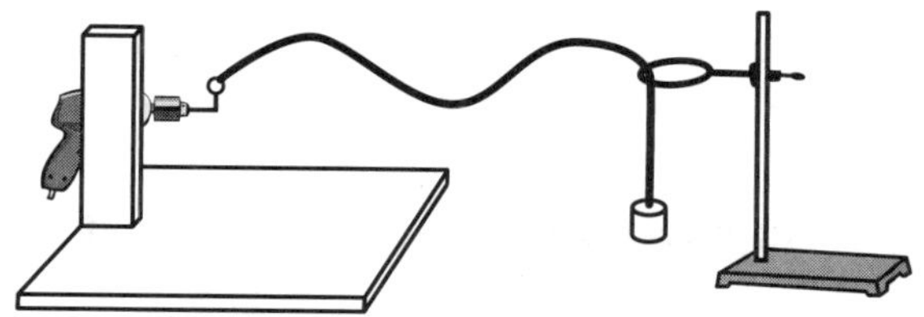

bolt for 1 minute and then immerse it into a container of cool water. Return the bolt to the vise and then temper the metal of the bolt by heating it gently for about 30 seconds. Allow the bolt to cool.

Create a stand for the drill by securing a 2″ × 4″ wooden board vertically along the center edge of a piece of plywood. The 2″ × 4″ must be oriented such that the 2″ side is facing the edge. Use three 3″ wood screws to secure the board to the plywood base.

Secure the eyebolt into the chuck of the drill. Use two pipe clamps to secure the handle of the drill to the vertical 2″ × 4″ with the drill oriented above the plywood base. Use two large C-clamps or adjustable clamps to secure the base to one edge of a table with the drill facing towards another table. Secure a small ring stand with a ring to the second table with the ring facing the drill assembly. Tie a 2.5-m fluorescent-colored nylon rope to the eyebolt. Drape the other end of the rope through the ring and tie a loop in this end of the rope about 30 cm below the ring. Hang a 1- or 2-kg mass from this loop.

Demonstration

Show students an individual wave pulse by plucking the rope near the drill to show how the pulse moves away from the drill, and then reflects off of the ring back towards the drill. Discuss how a transverse wave propagates at a right angle to the disturbance.

Make sure the drill and the rheostat are with both in the off position and then plug the drill into a rheostat transformer. Depress and lock the switch of the drill. Slowly adjust the speed of the drill using the rheostat until the string vibrates with a large single loop—this corresponds to the natural frequency of the string system. Adjust the speed of the drill to show that the loop appears only if the drill is tuned to a specific frequency. Slowly increase the speed of the drill until two distinct loops are present—this frequency corresponds to the second harmonic. While the string is vibrating, touch the node of the string to show that the energy still travels through the rope, emphasizing that touching the node does not alter the original wave form. Continue to increase the speed of the drill until more harmonics are displayed. The motion of the string can be "frozen" using a strobe light tuned to the frequency of the string. If the strobe includes a tachometer, one may show that the frequencies of the harmonics are whole number multiples of the natural or fundamental frequency.

Seeing Sound

Sound waves are made visible using a homemade device akin to a gas barbeque burner.

Application	Waveform • Standing waves • Amplitude • Frequency • Pitch • Nodes
Theory	As sound travels though a medium, the energy of the wave compresses and stretches the particles in the medium. In this demonstration sound is transmitted through natural gas or propane which is contained in a tube with a line of holes along its full length. When the tube is filled with natural gas, prepare and light. The sound pushes the gas through the holes with varying pressure corresponding to the low and high pressure regions of the wave. If the frequency of the sound is matched with the length of the tube, a standing wave will be produced. The greater the sound volume, the greater the pressure on the gas. Lighting the gas reveals the gas pressure difference with a corresponding large or small flame.
Materials	Auto engine sealant, high temperature, silicone-based

Drill, hole saw bit, 3″

Drill bit, ½″

Drill bit, ³⁄₆₄″ (#54)

Gas/liquid inlet, ½″ (available in the plumbing department at home hardware centers)

Match, long fireplace, or long barbeque lighter

Musician's electronic keyboard

Patch-cord, one end to the output of the keyboard, the other with two alligator clips

Plywood, 5″ × 5″ × ½″, 4 pieces

Rain downspout, 4′–10′ piece

Screwdriver

Screws, ¾″, 12

Speaker, 3″

Note: If the output of the keyboard is not sufficient to produce a wave form, an amplifier may be placed in between the speaker and keyboard (e.g., Mini Audio Amplifier from RadioShack, Catalog #277-1008).

Safety Precautions

Follow manufacturers' directions when using power tools. Do not use the flame tube for more than 5 minutes at a time to prevent damage to the speaker and scorching of the wood due to the build-up of heat. Make sure that the flame tube is placed firmly on a table so that it does not fall. Inspect the device for gas leaks prior to use. Keep away from flammable objects. Practice this demonstration before performing it in front of a class. Wear safety glasses and follow all laboratory safety guidelines.

Preparation

Drill a 3″ hole into the center of three of the four blocks of wood. Slide one of the blocks onto each end of a rain downspout. Use a silicone-based high temperature sealant to secure the blocks so they are flush with the ends of the pipe. Align the blocks so they support the tube in a horizontal position on a table. Use sealant to glue a 3″ speaker over the third block with the hole in it. Secure the speaker to this block using four ¾″ screws. Drill a ½″ hole into the center of the fourth block of wood. Use sealant to secure a ½″ gas inlet in this hole. Allow the sealant to set overnight. Use a permanent marker to draw a line along the top of the full length of the downspout. Use a ruler to place marks along this line every 2.5–5 cm away from either block. Drill a hole at each mark using ³⁄₆₄″ bit. Remove any metal burrs from the inside of the downspout. Use sealant and four screws to secure the board containing the gas inlet over one end of the downspout assembly. Attach the speaker board on the opposite end in a similar fashion. Use sealant liberally when assembling this device to prevent any gas leaks.

Demonstration

Use a patch cord to connect a musical keyboard to the speaker of the flame tube. Connect a gas source to the gas inlet of the flame tube. Set the keyboard to play a continuous tone. (Select a note that will produce a standing wave on your flame tube with at least five wave crests.) Play a note on the keyboard and ask students to describe what the sound "looks like." Turn the gas on and allow the tube to fill for a minute or until you first notice the odor of gas. Light the gas using a long fireplace match or long barbeque lighter. The variations of the pressure inside the tube due to the sound wave will produce a standing wave made visible by the flames. Adjust the volume control on the keyboard to see how the amplitude of the flames is affected. Vary the notes to see the various waveforms due to the change in pitch. If the keyboard has a demo mode, try playing a song into the tube. A radio or other source of music may be connected to the tube resulting in a visually exciting display of dancing flames.

Reference

Sutton, Richard. *Demonstration Experiments in Physics;* American Association of Physics Teachers: College Park, MD, 1938; p S-130.

Standing Waves of a Baseball Bat

The sweet spot of a bat is located.

Application

Standing waves • Nodes • Sports science

Theory

The most effective method of hitting a baseball with a bat is to have the ball connect with the bat at a spot known as the "sweet spot." Striking the ball at any other spot along the length of the bat will induce a standing wave, or vibration between the two nodes within which the bat oscillates. One node is located where the batter holds the bat, while the other node is located at the sweet spot or center of oscillation. The center of oscillation of the bat is also called the center of percussion. An object striking a bat at its center of percussion will encounter the transfer of most of the bat's kinetic energy. If collision occurs at other points along the bat, then less energy is transferred. Some of the energy is then transferred to the hands of the batter which result in a stinging sensation. The design of a baseball bat can have an effect on the location of the sweet spot. Bats made from aluminum are hollow, which gives rise to a larger sweet spot. Thus, aluminum bats are easier to hit with because the large sweet spots give the batter a larger target. Tennis rackets also exhibit a similar effect.

Materials

Baseball	Hot glue gun
Baseball bat, aluminum	Masking tape, roll
Baseball bat, wooden	Metal rod, $\frac{3}{16}''$, 50 cm
Bowling ball, 16-lb	Permanent marker
C-clamp	Ring stands, 2
Drill	String, 1.5 m
Drill bit, $\frac{1}{4}''$	Utility clamps, heavy-duty, 2
Hot glue	

Safety Precautions

To prevent injury and damage to the bat, volunteers should be instructed not to "wind up" before striking the ball with the bat. Students should also swing the bat in a direction away from the audience. Always follow laboratory safety rules while performing demonstrations.

Preparation

Beginning at the top of the bat, use a permanent marker to draw rings every 2 cm around the circumference of the wooden bat. Do the same with the aluminum bat.

Drill a $\frac{1}{4}''$ hole through the handle of a wooden baseball bat. Do the same for an aluminum bat. Pass a $\frac{3}{16}''$ metal rod through the hole in the wooden baseball bat and secure it to a ring stand as illustrated in the diagram. Clamp the apparatus to the edge of a table. The bat should be arranged so it can freely swing around the rod.

Use hot glue to secure a 1.5-m string to a baseball. Suspend the baseball from a fixed point above the bat.

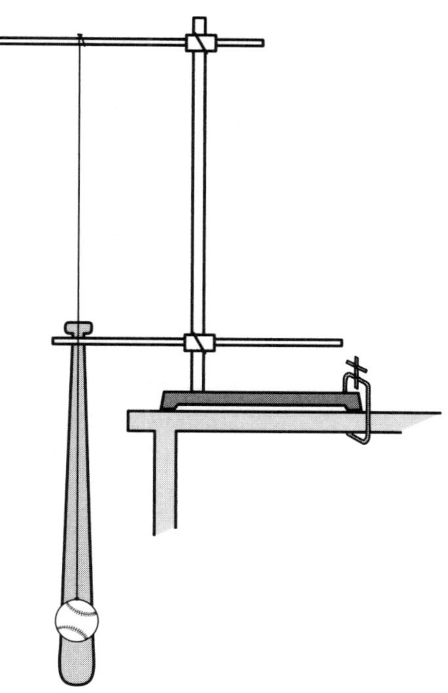

Demonstrations

Locating the Sweet Spot

Suspend the baseball so it is touching the tip of the wooden bat in its rest position (lowest point of swing). Raise the bat 90° and allow it to fall freely and strike the baseball. Note the height the baseball swings. Raise the baseball 2 cm and repeat the collisions. Keep repeating the trials until you are able to identify the spot on the bat that causes the ball to rise to the highest point—this is the sweet spot. Mark the sweet spot with a piece of masking tape.

Repeat the experiment using an aluminum bat. Students will notice that the sweet spot on the aluminum bat is larger.

Feeling the Sting

Balance a bowling ball on a roll of masking tape placed sideways near the edge of a sturdy table. Invite a student to take the aluminum bat and strike the bowling ball with the bat—ensuring that the two connect at the sweet spot. To prevent injury and damage to the bat, the volunteer should be instructed not to "wind up" before striking the ball. The student should also be swinging away from the audience. Have the student repeat the procedure but this time striking the ball at a point other than the sweet spot. Ask the student to describe what was felt in their hands in the two instances. Have the same student repeat the experiment using a wooden bat. There should be a pronounced difference when striking the bat on the sweet spot.

Disposal

Inspect the bats for damage after each demonstration. Once a bat cracks, it should be disposed of.

Good Vibrations

Application | Natural frequency • Pitch • Musical scales

Theory | When an object vibrates back and forth, the kinetic energy in the vibrating object is transferred to the surrounding air creating sound waves that travel to our ears. The vibrating air in turn transfers this energy to our eardrums. The brain recognizes these vibrations as sound. The frequency of the vibrations depends on the length of the vibrating object. This frequency is called the natural frequency as it is the lowest pitch that the object can emit. Generally, the longer the object, the slower the vibration and lower the pitch. The composition of the material itself also contributes to the frequency produced.

This demonstration uses a number of objects to illustrate the relationship between length and pitch. In particular, a series of copper pipes is tuned to a specific musical scale. The lengths of the pipes (L) may be determined using the following equation:

$$L = \sqrt{\frac{\pi v_L K m^2}{8 f_n}}$$

where v_L is the speed of sound in the material, K is the radius of gyration, m is a constant and f_n is the frequency of the sound. The speed of sound in copper is 3650 m/s. The radius of gyration, K, depends upon the inner and outer diameter of the tube:

$$K = \frac{\sqrt{(inner\ radius)^2 + (outer\ radius)^2}}{2}$$

The radius of gyration reflects the degree that the pipe wiggles back and forth (like a tuning fork) as it resonates. The constant m depends upon the mode of vibration, n. When $n = 1$, $m = 1.194$; when $n = 2$, $m = 2.988$; when $n = 3$, $m = 5$ . . . . $(2n-1)$.

When using ½″ M-gauge copper plumbing pipe, the lengths of pipe may be cut as follows:

Scale Degree	Do	Re	Mi	Fa	Sol	La	Ti	Do
Perceived Note	B	C#	D#	E	G	G#	A#	B
Frequency (Hz)	984	1125	1234	1320	1500	1695	1875	2000
Length (mm)	274	256	244	235	220	201	196	191

The lengths outlined above correspond to the frequency *perceived* and thus may not correspond to the equation described above. Human hearing is quite complex. For example, if frequencies of 800, 1000 and 1200 Hz are combined, most listeners will perceive a pure frequency of 200 Hz. Consequently, perceived pitch is not necessarily related to the fundamental frequency of a vibrating pipe. Further discussion on perception versus actual frequency may be found in the reference listed below.

The above scale uses the "philosophical" or scientific standard (C5 = 256 Hz) rather than the standard used by musicians when tuning their instruments (C5 = 261.6 Hz). The two standards are slightly different—enough to produce beat frequencies when played together. The "scientific" standard was developed so that the octaves of C (the fundamental frequency of C being 2 Hz) are powers of 2 (2^8 = 256). This is the typical frequency that is used in the classroom because it is a whole number and makes for easier calculations. Most tuning forks used in the science classroom will use this standard, C = 256 Hz. The standard "musical" middle C is 261.6 Hz. This is the frequency that is produced by the middle C string of a piano. This middle C frequency is based on setting the A note above middle C at exactly 440 Hz (at room temperature). The A note is 9 semitones over middle C. Each "semitone" is 1.05946 ($2^{1/12}$) times the previous semitone frequency.

Materials

Graduated cylinder, glass, 50-mL (preferably a tall-form)

Pipe, ½″ copper ("M" gauge)

Pipe cutter for ½″ copper pipes

Rubber mallet or rubber activator

Ruler, wooden or meter stick

Tuning fork

Wash bottle, filled with colored water

Optional: Adjustable strobe light

Safety Precautions

Although this demonstration does not pose any obvious risks, always follow laboratory safety rules.

Preparation

Use a pipe cutter to cut the copper pipes into the lengths outlined above. If necessary, use a file to trim the pipes if the perceived pitch is too low. Use a hammer to *slightly* flatten one end of a pipe that has a pitch that is too high.

Demonstrations

Metal Chimes

With students seated, drop the set of copper pipes onto the floor one pipe at a time from longest to shortest. Students will hear a musical scale of notes. Ask students if they can explain why each pipe has a unique sound. The frequency of the pipe depends upon its length—the shorter the pipe, the higher the frequency.

Vibrating Tuning Fork

Strike a tuning fork against a rubber mallet and then ask students if they can explain what the tuning fork were doing to produce the sound. Ask students what would happen if the tuning fork were dipped into a cup of water. The vibrating motion of the tines will demonstrate that the tuning fork is oscillating back and forth, causing the water to splash.

Freeze and Slow the Vibration (Optional)

Shine a strobe light onto the moving tines of a tuning fork. Adjust the flashes of the strobe to match the frequency of the tuning fork—this should result in the "freezing" of the motion of the tines. Adjust the frequency slightly up and down to view the apparent slowed motion of the tines.

Closed Tube Resonance

Ask students what will happen if you blow across the top of an empty 50-mL graduated cylinder. Most likely, students will expect to hear a whistle-like tone. Blow across the bottle to confirm their hypotheses. Next, ask what effect there would be on the sound if water were added to the cylinder and then the process repeated. As you blow across the top of the cylinder, squirt water into the cylinder using a water bottle. The pitch or frequency of the tone should increase as the length of the column of air in the cylinder decreases.

Vibrating Objects

Hold a wooden ruler or meter stick on the edge of a desk, akin to a diving board, and gently pluck the stick to produce a sound. Slowly pull the stick inward to shorten the vibrating end of the stick to produce a series of higher frequencies.

Ask students to summarize how the musical tones were produced using these four objects.

Play a short song such as "Mary Had a Little Lamb" or "Do Re Mi"— a tune from "The Sound of Music" (Doe a Deer)—to illustrate the relationship between musical notes and the length of the ruler.

References

Thanks to Brad McCabe, former Staff Scientist, Flinn Scientific Inc. for his explanation regarding the difference between the scientific and musical scales.

Forinash, Kyle, et al. "Wind Chime Physics"; *American Journal of Physics;* 1990; vol 58, January, p 82.

A metal rod is stroked causing it to vibrate, producing a loud sound.

Application | Resonance • Standing waves • Harmonics • Amplitude • Pitch

Theory | In this demonstration a metal rod is stroked with tacky fingers, like a bow stroking a violin string. The sound produced is due to the stretching of the metal and its rebound or vibration—similar to the rebound of a spring. As the rod vibrates, energy is transferred to the surrounding air, causing the air molecules to vibrate in a similar fashion. In turn, the air transfers this vibration to your ears. The method used to vibrate the rod in this demonstration works only if the rod is held in certain positions corresponding to the nodal points of the rod. If held at its center, the rod will vibrate at the second harmonic or twice its natural frequency. If held at one-third of its length, the rod vibrates at the third harmonic, or three times the fundamental frequency. It is worth mentioning that a discrepancy exists in the common terminology used to describe the harmonics. Although scientists use the terms fundamental frequency and first harmonic synonymously, some musicians, tuners, and piano makers define the first harmonic as twice the frequency of the fundamental. In this demonstration a Styrofoam® cup is pressed against the end of the rod as it "sings" in order to amplify the sound. The amplification is due to the coupling effect between the cup and the rod. The cup provides a larger surface area in contact with the air. In short, the cup is shaking more air and producing a louder sound.

Materials |

Aluminum rod, ⅜″, 60–100 cm	Pipe cutter
Cup hook	Styrofoam® cup
Drill	Tape, electrical, 3 different colors
Drill bit, ⅛″	Violin rosin
Hammer	Ziploc® freezer bag, small
Meter stick	

Safety Precautions | Although this demonstration does not pose any obvious risks, always follow laboratory safety rules. Follow manufacturers' directions when using power tools. Wash hands thoroughly with soap and water.

Preparation | Drill a ⅛″ hole in the center of one end of the aluminum rod. Insert a cup hook into this hole. Use construction adhesive to secure the hook, if needed. Use a meter stick to mark the ½, ⅓ and ¼ lengths of both ends of the rod (5 marks). Use a pipe cutter to score a ring around the rod at each of these locations. Wrap a piece of colored electrical tape to mark the locations on one-half of the center mark, leaving the other half free of tape.

Place a piece of violin rosin into a plastic freezer bag (heavy-duty plastic bag) and pulverize the rosin into a powder using a hammer.

Demonstration | Coat your thumbs and forefinger with powdered rosin. Hold the center of the aluminum rod tightly with two fingers of your other hand and stroke the rod firmly with your rosin-coated fingers to make the rod "sing." The stroking action is akin to stroking a violin string with a bow—it takes practice. As the metal rod is stroked, the amplitude of the sound will steadily increase with each stroke until the sound is nearly painful to your ears. Ask students why sound is emanating from the rod. Discuss how the rod is being stretched, which causes it to vibrate back and forth like a spring. Each time the wave pulse reaches the ends of the rod, energy is transferred to the surrounding air, causing the air molecules to vibrate also. The vibrating air molecules in turn transfer the energy to your ears.

Hold the rod and stroke it at each of the different marked positions to observe the different frequencies produced. Ask students to explain why the different frequencies are dependent upon where the rod is held. Have students describe what each wave form looks like as the various frequencies are produced. Hold the rod slightly off the center mark and attempt to make it "sing." Discuss how where the rod is held dictates which harmonic is produced.

Ask students how the rod can be made to vibrate at its first harmonic (also called the fundamental frequency). Show them that the fundamental frequency may be achieved by holding the rod by its end—the nodal location for the first harmonic.

Finally, stroke the rod while holding it at its center until a piercing sound is heard. While the rod is still vibrating, press the bottom of a Styrofoam cup against one end of the vibrating rod. The cup will amplify the sound greatly.

You Can't Beat a Singing Rod

Two vibrating rods interfere with one another and produce a beat frequency.

Application | Interference • Beats • Resonance • Pitch

Theory

In this demonstration a metal rod is stroked with tacky fingers like a bow stroking a violin string. The sound produced is due to the stretching of the metal and its rebound or vibration, similar to the rebound of a spring. As the rod vibrates, energy is transferred to the surrounding air causing the air molecules to vibrate in a similar fashion. In turn, the air transfers this vibration to your ears. When two sources of sound emit the same frequency, the amplitude or loudness of the sound increases since the waves interfere with one another constructively. These two frequencies are said to be "in phase" since their waveforms are identical with respect to time. If the sources produce different frequencies, then the waves will interfere both constructively and destructively—going in and out of phase. The rate at which they go in and out of phase is known as the beat frequency, in which the amplitude of the sound will oscillate up and down. The frequency, f, of the beat is dependent upon the difference between the frequencies of the two interfering sources: $f_{beat} = f_1 - f_2$. The beat frequency is most noticeable when the difference between frequencies is small, such as 5 Hz.

Materials

Ring stands, 2, or overhead hooks

Singing rods, 2, one being 1-cm shorter than the first (see Singing Rods, page 219).

Thread

Violin rosin

Safety Precautions

Although this demonstration does not pose any obvious risks, always follow laboratory safety rules. Wash hands thoroughly with soap and water.

Preparation

Suspend two loops of thread from overhead hooks or two ring stands. These will be used to hang the vibrating rods.

Demonstration

Coat your thumb and forefinger with powdered rosin. Hold the center of one rod tightly with two fingers of your other hand and stroke the rod firmly with your rosin-coated fingers to make the rod "sing." The stroking action is akin to stroking a violin string with a bow—it takes practice. As the rod is stroked, the amplitude of the sound will steadily increase with each stroke until the sound is nearly painful to your ears. Set this rod aside. Stroke a second but slightly shorter rod and ask students if the frequency of the second rod matches the first. Most likely they will believe the two rods emit the same frequency since they are nearly identical. Stroke the rods one at a time and suspend the vibrating rods from loops of thread suspended from the overhead hooks or from separate ring stands. Have the students listen to the rods vibrating at the same time. They should notice a beat frequency with the amplitude rising and falling due to the interfering frequencies of the two rods.

Shifting Sound

The pitch of a buzzer is changed by twirling it in a circle.

Application | Doppler effect • Frequency change • Pitch • Relative velocity

Theory | When the difference in velocity between the source of sound and the observer is any value other than zero, the pitch of the sound will change. This is known as the Doppler effect. A common example of this is observed when a car blaring its horn approaches and then passes an observer. As the car approaches the observer, the pitch of the horn is higher because the velocity of the car is added to the speed of sound, causing the sound wave to travel relatively faster than if the source were stationary. Increasing the relative velocity of the sound increases its pitch. After the car passes the observer, the pitch decreases since the movement of the source causes a decrease in relative velocity of the sound wave moving toward the observer. This effect is apparent whether the source of sound is moving toward the observer or whether the observer is moving toward the source of sound. The Doppler effect occurs in any type of wave including electromagnetic waves. Police radar units are based on the Doppler effect by calculating the relative velocity of the returning microwave pulse that was emitted toward the vehicle. Weather forecasters use similar techniques to determine the speeds of weather formations. Astronomers use the blue–red shifting of light to determine the relative movement of astronomical objects with respect to the Earth.

Materials | Battery clip, 9 V

Buzzer, 9-DCV (available at RadioShack, e.g., RS Model 273-055)

Knife

Mason line or nylon cord, 1.5 m

Pencil or knitting needle

Soft foam ball, 6″–8″ (e.g., Nerf® soccer ball)

Soldering iron and solder

STSP (Single Throw Single Pole) switch (available at RadioShack, e.g., RS Model 275-327)

Tape, duct

Wire, 18–24 gauge, $\approx$ 10 cm long

Wire strippers

Safety Precautions | Twirl the Doppler device in an area free of obstacles, such as suspended light fixtures or walls. Review manufacturers' directions before using a soldering iron. Always follow laboratory safety rules.

Allow the soldering iron to heat up. Strip 2 cm of the wire insulation from one end of the wire. Connect the leads of the buzzer, the battery clip wires, the switch, and the wire in a single continuous loop and then solder together the four connections.

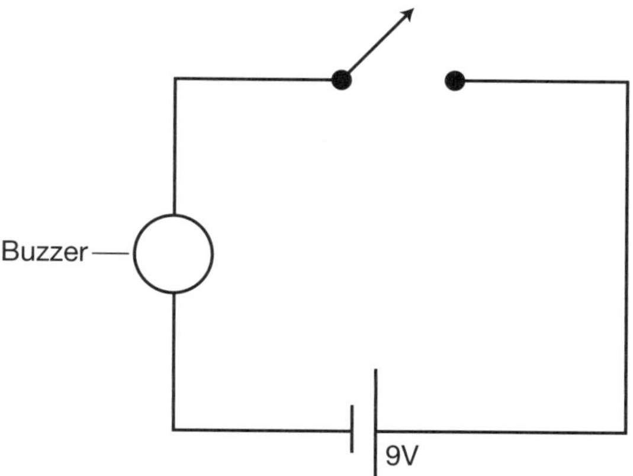

Use a knife to cut a small slit into the foam ball large enough to insert the buzzer assembly. Attach a piece of nylon cord to a pencil or knitting needle using a piece of duct tape and then use the pencil to thread half of the cord through the center of the slit to the other side of the ball. Tie the opposite end of the cord to the battery and then wrap a piece of duct tape around the string and battery to secure the string firmly. Attach the battery to the buzzer assembly and insert it into the slit of the ball. Arrange the switch so that it is easily accessible through the slit. Detach the cord from the pencil and then tie a safety loop large enough for your hand to slip through it. Pull the cord until the buzzer assembly is snug inside the ball.

Turn the buzzer on and ask students to notice the pitch of the sound. Slip your hand through the safety loop of the assembly and twirl the ball above your head in a horizontal plane. Ask students to describe what they hear. As the ball approaches the students, the pitch should be higher than when the buzzer is stationary. As the ball moves away from the students, the pitch should be lower. The faster the ball is twirled, the greater the change in pitch. Ask students if they can predict what the demonstrator hears. Invite a student to verify the prediction by having him or her perform the demonstration.

Notes

Chapter 12

Thermal Physics

Mechanical Equivalent of Heat

A device is used to show that mechanical potential energy is converted into heat energy.

Application	First Law of Thermodynamics • Specific heat capacity • Conservation of energy

Theory
According to the First Law of Thermodynamics, heat may be added to a system when work is done on that system. In the first part of this demonstration, the gravitational potential energy of a hammer is transferred to a block of wood, causing its temperature to rise. The temperature increase is measured using a heat sensitive liquid crystal film. The second part of the demonstration uses a device to quantitatively illustrate the amount of heat energy gained from a transfer of potential energy. The device contains a known amount of metal shot that is dropped inside a tube of known length. The dropping is repeated 100 times so that an appreciable difference in temperature may be observed. The mechanical equivalence of heat may be summarized by:

$$\Delta PE_g = \Delta Q$$
$$m gh = m \Delta T C_p$$

where $h = 100 \times$ length of the pipe, and C_p is the specific heat of the material.

Materials
Drill

Drill bit, $\frac{1}{8}''$

Hammer

Hardwood block (e.g., $1'' \times 3'' \times 3''$ maple or oak)

Lead shot, 500 g

Liquid crystal film, heat-sensitive sheet

PVC cement

PVC end caps, 2

PVC pipe, $1\frac{1}{2}''$, 50 cm long

Tape, electrical

Temperature probe thermometer, digital

Safety Precautions
Although this demonstration does not pose any obvious risks, always follow laboratory safety rules.

Preparation
Some heat-sensitive films come with an insulating foam backing; remove this prior to the demonstration. Drill a $\frac{1}{8}''$ hole into one PVC end cap. The hole should be just large enough to insert the digital temperature probe. Use PVC cement to attach the end cap to the PVC tube (pipe). Pour 500 g of lead shot into the tube and attach the second end cap using PVC cement. Cover the hole in the end cap with tape to prevent any of the shot from escaping.

Demonstration | Ask students what they expect will happen to a piece of wood—besides some physical deformation—if the wood is struck repeatedly with a hammer. Discuss how the potential energy of the hammer will transfer energy to the block of wood as it is struck, causing the temperature of the wood to rise. Hold a heat-sensitive liquid crystal film on the surface of the block to verify its initial relative temperature. Strike the wood block with a series of blows. Once more, hold the liquid crystal film on the surface. The film should change color over the area that was struck.

Show students the PVC Mechanical Equivalence of Heat Demonstrator and explain how it may be used to quantitatively demonstrate how mechanical potential energy may be converted into heat energy. Measure the initial temperature of the shot by inserting a temperature probe through the hole in the end cap. Remove the thermometer and cover the hole with tape again. Slowly rotate the device 100 times, allowing the shot to fall the full length of the tube each time. Measure and record the final temperature of the shot. Compare the change in potential energy of the shot with the change in the heat of the wood.

See It Grow

A technique is used to magnify the subtle change in the length of a heated metal rod.

Application	Thermal expansion • Kinetic theory
Theory	A candle is used to heat a metal rod to show how most materials expand upon heating. This demonstration uses a laser beam and rotating mirror assembly to magnify a small movement similar to the apparatus used in the Lazy Walls demonstration, p 6. The technique employs a mirror mounted on a pin with the metal rod placed on the pin. The rod expands as it is heated, causing the pin to rotate. A laser beam focused on the mirror will reflect a beam that will deflect, dramatically magnifying the expansion of the rod. Engineers must take thermal expansion into account in their designs. For example, bridges and sidewalks include expansion joints to allow for seasonal temperature changes to prevent buckling or collapse of the structure. The expansion and contraction of the wires in response to temperature changes must be considered when deciding the degree of sag when suspending overhead cables. In contrast to most materials, rubber will contract upon heating because it has a negative thermal coefficient of linear expansion. This property is beneficial as it improves the grip of automobile tires as they heat up during use. Water has a negative coefficient in the range of 0–4 °C, which explains why ice is less dense than liquid water. If water did not have this property, life as we know it would not exist.
Materials	Adhesive, construction or epoxy Aluminum rod, 24″ × ³⁄₁₆″ (available at home hardware centers) Beakers, 400- or 600-mL, 2 Candle and holder Laser pointer Matches Mirror, small (e.g., 1″ round) Pin, large *Optional:* Rubber band, a small mass, ring stand, and blow dryer.
Safety Precautions	Do not direct a laser beam into eyes. A laser produces an intense beam of light that may cause permanent damage if directed toward an eye. Follow all laboratory safety guidelines. Wear safety glasses whenever working with heat or glassware.
Preparation	Use construction adhesive or epoxy to glue a large pin to the back of a mirror.
Demonstration	Hold up the aluminum rod and ask students what would happen to the rod if it were heated a few hundred degrees, but still below its melting point. Discuss how most objects expand upon heating. Ask students to propose how the linear expansion of

the rod may be measured as it is heated. Lay an aluminum rod on two inverted beakers situated at either end of the rod. Slide the pin of the mirror–pin apparatus between the rod and one of the beakers. Reflect a laser beam off the mirror onto the ceiling or wall. Show students how the mirror–pin assembly rotates as the rod moves in the direction of its length. Center a candle under the rod so that the tip of the flame is touching the rod. Direct students' attention to the fact that the laser dot moves across the ceiling or wall as the expansion of the rod is magnified by the mirror–pin apparatus.

Optional: Ask students what they expect will happened to a stretched rubber band when it is heated. Suspend a mass from a thick rubber band attached to a ring stand and heat the band using a blow dryer. To the students' surprise, the band will contract.

Boiling Water in a Balloon

A candle is used to heat water contained in a balloon.

Application | Specific heat capacity • Conduction

Theory | Specific heat capacity refers to the amount of heat energy a substance can absorb to cause a rise in temperature—measured in Joules/kg·°C. For example, 1 kg of water requires 4186 Joules of energy to raise its temperature by 1 °C. In this demonstration an air-filled balloon is brought into contact with an open flame. The balloon immediately bursts because the heat energy supplied by the flame raises the temperature of the rubber, causing it to burn. In contrast, if the balloon is filled with water, the temperature of the balloon cannot rise above the boiling point of the water, which is lower than the flash point of the rubber. The water serves as a heat sink, pulling the heat energy away from the rubber. Bodies of water on our planet regulate local temperatures by absorbing heat on a hot day and radiating heat on cold day. Consequently, temperatures along bodies of water are slower to change and are not as extreme. In contrast, temperature changes in deserts are very extreme because there are no bodies of water to regulate the temperature.

Materials | Balloons, 8″–12″ round, 2

Candle and holder

Matches

Ring stand and ring

String, 30 cm

Safety Precautions | Although this demonstration does not pose any obvious risks, always follow laboratory safety rules. Wear safety glasses whenever working with heat or glassware.

Demonstration | Light the candle. Fill a balloon with air and ask students what will happen if the balloon comes in contact with the flame. Upon discussion bring the balloon in contact with the flame to show that the balloon will absorb the heat, causing it to burst. Take a second balloon and fill it with about 500 mL of water, then tie off the end and use a short piece of string to suspend the balloon from a ring stand. Ask students what will happen if the candle is placed in contact with this balloon. On concluding the discussion, bring the balloon in contact with the flame to show that the balloon will not burst this time. Discuss how the water serves as a heat sink and prevents the temperature of the balloon from rising above its flash point.

The thermodynamic properties of a commercial toy known as a "drinking bird" are explored.

Applications

Second Law of Thermodynamics • Carnot engine
Evaporative cooling • Ideal gas law

Theory

The toy "drinking bird" is an example of a Carnot engine. The felt-coated head of the bird is first immersed in water to initiate the process. As the water evaporates from the felt, the temperature of the head cools, subsequently causing the vapor inside the head to contract according to the ideal gas law. The head serves as the cold reservoir of the heat engine while the body serves as the hot reservoir, since the body is warmer than the head. The pressure difference between the head and the body causes the fluid (methylene chloride or similar volatile liquid) in the body to be drawn up toward the head. This changes the center of mass of the bird, causing the head of the bird to tilt into the cup of water. With the neck in the horizontal position, the seal separating the body and head is broken, causing the fluid to return to the body and subsequently raising the neck of the bird. The entire process then repeats itself, following the steps of the Carnot engine. The process will continue until the head can no longer replenish itself with water from the cup. If the body of the bird is initially cooled in a cup of ice, the bird will not start drinking until it has warmed up. This is due to the fact the vapor pressure inside the head is much greater than the vapor pressure in the body and the evaporative cooling of the head is not sufficient to cool the head. Warming the water in the cup will cause the drinking rate to increase due to the increased rate of evaporation.

Materials

Commercial drinking bird toys (available from science suppliers or novelty stores), 2

Cups, plastic, drinking glasses or beakers, 2

Ice, 1 cup

Water, distilled, 2 cups, one at room temperature and the other hot

Note: Serve the bird distilled water in order to prevent mineral deposits on the felt.

Safety Precautions

Although this demonstration does not pose any obvious risks, always follow laboratory safety rules.

Preparation

Place one drinking bird on ice for at least 15 minutes.

Demonstration

Immerse the head of a drinking bird (the one at room temperature) into a cup of distilled water—both at room temperature. Adjust the drinking bird such that it will drink out of a cup of water. Once the bird begins to "drink," ask the class to explain how the bird operates. Direct students' attention to the neck of the bird and how the fluid moves from its body toward its head. Emphasize how the water on the felt evaporates causing the head to cool and the vapor inside the head to contract, creating a pressure difference between the head and the body. As a result, the fluid inside the body is drawn toward the head of the bird. When the fluid moves up the neck, the bird's center of gravity shifts, causing the head to tilt downward. This allows the bird to take another "drink" and again wet its head, causing the entire process to repeat itself. Discuss the process in terms of heat engines and how the head of the bird serves as the cold reservoir and the body as the hot reservoir of the engine.

Bring out the second drinking bird that has been previously cooled on ice. Ask students if "this bird will be as thirsty as the first bird." Activate the second bird by wetting the felt with water. Discuss how temperature affects the vapor pressure of the fluid inside the second bird and how the lower temperature slows the drinking process similar to the Carnot cycle.

Reference

Vemulapalli, G. K. "A Discourse on the Drinking Bird"; *Journal of Chemical Education;* American Chemical Society: Washington, DC, 1990; vol 67, Number 6, p 457.

Refrigerants Are Cool Gases

Applications Second Law of Thermodynamics • Adiabatic expansion and compression
Ideal gas law

Theory A gas can be liquefied if it is cooled below its boiling point or if it is below its critical temperature and under sufficient pressure. It is common to see gases such as butane and propane in small containers around the home—these gases can be liquefied and kept in relatively low-pressure containers. Natural gas (methane) is harder to liquefy and would require bulkier and less economical containers. The critical temperature is the temperature below which a gas can be liquefied (condensed) by increasing the pressure. Liquid Petroleum Gases include propane (critical temperature 97 °C) and butane (critical temperature 152 °C), both of which are gases at room temperature and pressure. They can be easily condensed to liquids at room temperature by increasing the pressure because their critical temperatures are above room temperature. In contrast, methane has a critical temperature of –82.7 °C.

The compression of gases is the first step in refrigeration. Air conditioners and refrigerators operate on the principle of the Joule–Thomson (Kelvin) effect, which states that when a real gas is allowed to expand rapidly, its temperature drops. In this case, no heat is transferred to or from the gas and no external work is extracted. The gas expands at a constant internal energy in accordance with the Second Law of Thermodynamics. In refrigeration and most household air conditioners, a refrigerant gas is initially compressed and liquefied, causing its temperature to increase. Liquefaction takes place in a condenser that is located outside the house. The liquid is then cooled, removing much of its heat energy. The cooled liquid is allowed to rapidly expand to a gas, thus lowering its temperature a great deal more. Adiabatic expansion of the gas takes place in an evaporator that is placed inside the house. The two systems are connected by tubing.

Materials Butane, canister (e.g., Ronson® butane)

Drill

Drill bit, ½″

Syringe, 50-mL

Syringe cap

Video camera and projector

Wood, block (e.g., 2″ × 4″ × 4″)

Optional: Thermocouple and multimeter.

Safety Precautions	Butane is highly flammable and a fire risk. Wear chemical splash goggles when conducting this demonstration. Follow manufacturers' safety guidelines whenever using power tools. Always follow laboratory safety guidelines.

Safety Precautions

Butane is highly flammable and a fire risk. Wear chemical splash goggles when conducting this demonstration. Follow manufacturers' safety guidelines whenever using power tools. Always follow laboratory safety guidelines.

Preparation

Use a ½″ drill bit to drill a ½ – ¾″ hole into the center of one face of the block of wood. Press the tip of the butane cartridge against the opening of the syringe, causing it to fill with the gas. Seal the syringe using a syringe cap. Insert the capped end of the syringe into the hole in the block of wood. The block will serve as a base to help keep the syringe upright as the gas is compressed.

Optional: The temperature rise/fall during compression/expansion of the butane may be visualized by inserting a thermocouple through a hole drilled into the side of the syringe near the tip. The thermocouple must be permanently sealed using plumber's epoxy or construction adhesive. Certain multimeters are equipped to provide a temperature reading when connected to a thermocouple. The thermocouple is relatively inexpensive and available from electronics catalogs. The thermocouple may be simply connected to a standard multimeter and the temperature is read in relative terms—as the voltage increases, the temperature increases.

Demonstration

Have a student inspect the syringe to verify that it indeed contains no liquid. Use a video camera to project a close-up of the syringe barrel—about the 0–10 mL area. Compress the gas slowly by pressing on the syringe. As the syringe is compressed, students will see droplets of liquid form near its bottom. If the syringe is slowly released, then the droplets will seem to disappear due to evaporation.

Reference

Bilash, B.; Gross, G.; Koob, J. *A Demo a Day—A Year of Chemical Demonstrations;* Flinn Scientific: Batavia, IL, 1998; vol 2, p 142.

Pressing the Temperature

As air is compressed and expanded, it is accompanied by a change in temperature.

Application

Ideal gas law • Second Law of Thermodynamics • Adiabatic expansion and compression

Theory

As air is pumped into a tire (of fixed volume), the temperature of the gas in the tire increases. Adding gas to the tire increases its pressure, which in turn causes a rise in temperature. This is summarized by the ideal gas law: $PV = nRT$. In this demonstration air is pumped into a bottle containing a liquid crystal thermometer. The thermometer registers the change in temperature that occurs as the pressure inside the bottle changes. If the bottle is pressurized quickly, the process does not transfer heat energy through the walls of its container. This is called adiabatic compression. If the pressure of the gas is decreased rapidly, the process results in a decrease in temperature, which is called adiabatic expansion. Adiabatic expansion is the key step in refrigeration during which a compressed cooling gas is allowed to suddenly expand inside a coil of tubing, causing the temperature of the tubing to decrease. Air drawn across the tubing may then be cooled.

Materials

Fizzkeeper® soft drink bottle pump

Liquid crystal fish tank thermometer

Soft drink bottle, plastic, 500-mL, with label removed

Optional: Salt water

Safety Precautions

Although this demonstration does not pose any obvious risks, always follow laboratory safety rules.

Demonstration

Tightly screw the Fizzkeeper onto a dry plastic soft drink bottle and begin pumping air into the bottle. Ask students what is happening inside the bottle as the air is pumped in. Following discussion, open the bottle and insert a liquid crystal thermometer, and then pump up the bottle once again. Have students compare the temperature of the air before and after the bottle is pressurized. Finally, release the contents of the bottle by rapidly removing the Fizzkeeper and observe the decrease in temperature of the air in the bottle.

Optional: If the bottle is flushed with salt water, leaving the contents with a high level of humidity, then the apparatus may be used to simulate the formation of a cloud. Clouds are formed in the atmosphere when relatively humid air undergoes a decrease in pressure. In this case increasing the pressure of the humid air and then quickly releasing the pressure will result in the formation of a cloud in the bottle. The salt ions suspended in the water vapor act as seeds to facilitate cloud formation, similar to the role of dust or other particulate matter in the air.

Entropy Card Trick

The odds of dealing a royal flush are used to illustrate the concept of entropy.

Application | Entropy • Spontaneous reactions • Disorder

Theory | The probability that a poker hand will contain five specific cards is the same, regardless of which five cards are specified. Thus there is an equal probability of dealing 10♥ J♥ Q♥ K♥ A♥ as dealing a hand containing 2♣ 4♥ 10♥ 8♣ J♦. However the first hand, a royal flush, strikes us as more highly ordered than the second hand, a "nothing" hand. The reason for this is clear. There are only 4 poker hands that are in the "state" of a royal flush; in contrast there are over 1.3 million "nothing" hands. The "nothing" state has a higher degree of disorder than the royal flush state because there are so many more arrangements of cards that correspond to the "nothing" state.

We can use the same reasoning for chemical systems. For example, consider two states for the gas molecules inside the following containers:

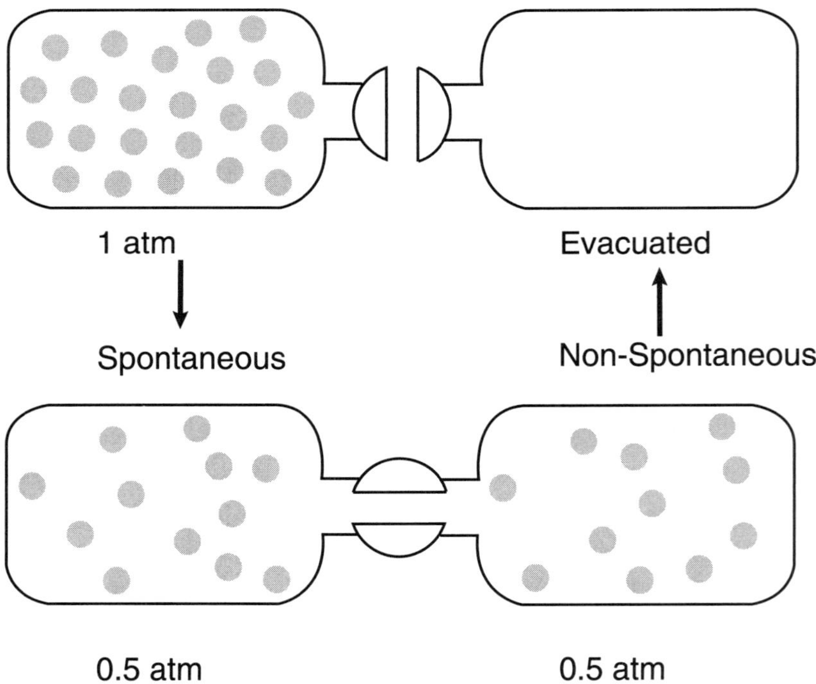

When the stopcock between the two chambers is opened, the volume available to the gas atoms is doubled. There are more possible arrangements of the atoms in the second figure than in the first; thus the randomness of the latter is greater than the former.

Materials	A deck of cards.
Safety Precautions	Although this demonstration does not pose any obvious risks, always follow laboratory safety rules.
Preparation	"Rig" the deck of cards so that when dealt, the five cards in each hand come out as indicated.
Demonstration	Deal two hands—face up:

Deal two hands—face up:

    10-♥, Jack-♥, Queen-♥, King-♥, Ace-♥

    5 random cards

Discuss the odds of obtaining a royal flush in terms of entropy. Repeat several times dealing two or more hands.

Notes

Chapter 13

Electric Forces and Fields

Charge It Up

A balloon is charged using three methods: conduction, polarization and induction.

Application   Static electricity • Charge production • Conduction • Polarization

Theory   When two objects made of different materials are rubbed against one another, an electric charge is collected on the surfaces between the two objects. If one of the materials is composed of atoms with a high affinity of electrons, then it will strip electrons from the atoms of the other material. As a result, the material with more electrons develops a slight negative charge while the other material, having had some electrons removed from it, develops a slight positive charge. This electric charge is called static electricity. A wool sweater placed in a dryer develops a positive charge because electrons are stripped from it as it tumbles inside the dryer. An acrylic rod can be positively charged by rubbing it with a woolen cloth. Since the rod has had some of its electrons stripped, it acquires a positive charge. In contrast, a PVC pipe will develop a negative charge when rubbed with a woolen cloth. Charging by conduction results in a lasting like-charge created in the originally neutral object.

In this demonstration, rubbing a balloon ion hair or wool gives it a net negative charge, allowing it to adhere to a neutrally charged surface in a process called polarization. As the balloon is brought near the neutral surface, the charge on the balloon induces the electrons in the atoms on the neutral surface to move away from the area near the balloon, giving the surface a net positive charge. As a result, there is enough attractive force to keep the balloon stuck to the surface. This effect, which is called charging by polarization, is also evident when a charged balloon is brought near a suspended neutral balloon. Although the suspended balloon remains electrically neutral, there is an attractive force between the two balloons. The induced charge is temporary as it is dependant upon the proximity of the charged object. Once removed, the polarized surface becomes neutral once again. Note that the polarized object remains electrically neutral—the electrons are simply redistributed resulting in the objects developing positive and neutral poles.

When a charged object comes into contact with another object of neutral or different charge, the charges redistribute themselves according to their affinity for electrons. The distribution is also dependant upon the surface area of the objects. The larger the surface area, the more room for electrons. In this demonstration a negatively charged balloon touches a neutral balloon, resulting in some of the electrons moving from the charged balloon to the neutral balloon. The balloons, now having a like charge, will repel one another. This process is called charging by conduction.

A polarized balloon can be made to have a lasting charge by grounding it. This is done by the balloon if the other surface, in this case a Mylar balloon, has a conduit for the displaced electrons. An electrically conductive Mylar balloon must be used in this demonstration in order to facilitate the flow of charge to and from the balloon. The nonconductive surface of the latex balloon would not work in this part of the demonstration. This process is known as charging by induction. The Mylar balloon is first polarized with the hand-held charged balloon, then the ground is removed, and finally, the hand-held charged balloon is removed. The process results in the Mylar balloon acquiring a net charge which is opposite to the charge of the hand-held balloon.

Materials

Acrylic rod, Lucite® or Plexiglas®, ½″ diameter, 40–50 cm long

Balloon, 8″ latex

Balloon, Mylar®, large

Balloons, 15″ latex, 2

Thread

Wool cloth or piece of fur

Safety Precautions

Although this demonstration does not pose any obvious risks, always follow laboratory safety rules.

Demonstration

Fill a small 8″ latex balloon with air, press it against a wall, and then release it. The balloon should fall to the floor. Next, rub the balloon against your hair or a wool cloth and press the balloon against the wall once again. The balloon should stick to the wall. Ask students to explain why the balloon does not fall this time.

Fill a large latex balloon with air and suspend it from a 1 m long thread hanging from overhead. Fill a second balloon with air and bring it near the suspended balloon. There should be no effect of one balloon upon the other. Next, rub the balloon against your hair or a wool cloth and bring it close to the suspended balloon, but do not allow the two balloons to touch. The suspended balloon will be drawn toward the rubbed balloon, demonstrating the presence of an attractive force. Discuss how the charge was produced on the hand-held balloon and why the electrically neutral balloon was attracted to the charged balloon.

Recharge the hand-held balloon and use it to touch the suspended balloon, rolling the surfaces together. The two balloons will now exhibit a repulsive force when held next to each other. Discuss how like charges repel one another.

Charge the suspended balloon with your hair or a wool cloth. Then charge an acrylic rod with the same cloth and bring it near the suspended charged balloon. The balloon will be attracted to the rod, demonstrating that the balloon and rod have opposite charges. Discuss how the type of charge produced (positive versus negative) is dependent upon the combination of the materials rubbed together.

Finally, fill a large Mylar balloon with air and suspend it from a 1 m long thread hanging from overhead. Charge a latex balloon by rubbing it on your hair or a wool cloth and then bring the latex balloon near the Mylar balloon (again, be careful the balloons do not touch). As with the neutral latex balloon, the Mylar balloon will be attracted to the charged hand-held balloon. With the latex balloon remaining near the Mylar balloon, touch the backside of the Mylar balloon with your free hand and then remove it. Now move the hand-held balloon away from the Mylar balloon. Charge an acrylic rod with the wool cloth and bring it near the suspended charged Mylar balloon. The Mylar balloon will be repelled, demonstrating that the balloon and rod have like charges. Discuss how grounding the Mylar balloon left it with a net charge.

A roll of tape is used as a charge indicator to differentiate between positive and negative static electric charge.

Application | Static electricity • Charge • Triboelectric series

Theory | Benjamin Franklin is credited with the realization that static electric charge was either the abundance or absence of a charge carrier. With the discovery of the electron still more than a century in the future at this time, Franklin decided to arbitrarily name these opposite charges "positive" and "negative"—perhaps an unfortunate choice. If the existence of the electron had been known at the time, it is likely that the charge carrier would have been called "positive."

The triboelectric series summarizes the affinity that a material has for electrons in relative terms from highest to lowest. Some of the more common materials in the classroom include: rabbit fur, glass, wool, nylon, rayon (viscose), silk, human skin, cotton, wood, Styrofoam® (polystyrene), acrylic, polyester, and latex (rubber) balloons.

Materials | Acrylic rod, Lucite® or Plexiglas®, ½″ diameter, 40–50 cm long

Dowel, wooden (e.g., ½″ × 50- cm)

PVC pipe, ½″ diameter × 40–50 cm long

Right angle clamp

Ring stand

Tape, transparent (Scotch® brand Matte Finish Magic™ tape works well)

Wool cloth or piece of fur

Demonstration | Create an electroscope-like device by draping a freshly dispensed, 30-cm piece of Scotch brand Matte Finish Magic tape in the shape of an upside-down "V" over a wooden dowel, which has been mounted horizontally from a ring stand. Direct students' attention to the fact that the two sides of the hanging tape repel one another, indicating that the tape is charged. Ask students if they can determine the type of charge on the tape. Following discussion, negatively charge a PVC rod by rubbing it with a wool cloth. Bring the rod toward the bottom of the leaves of tape and then slowly raise the rod between the tape. As the rod rises, the leaves of tape will part, indicating a force of repulsion. Students may conclude that the leaves and rod have the same charge. Repeat the demonstration using a positively-charged acrylic rod.

Recreate the electroscope with a new piece of tape. (The tape needs to be periodically replaced since the charge dissipates within minutes.) Take a 10-cm strip of tape and rub it against the surface of a table. Remove the tape and hold it next to the leaves of the electroscope to determine its relative charge. Repeat with other tape–surface combinations.

Reference   Mellan, Walter Roy. "Inexpensive Electrostatic Halos"; *The Physics Teacher;* American Association of Physics Teachers, 1990; vol 28, p 612.

The force of attraction between two charged objects is dramatically demonstrated by moving a large piece of wood with static electricity.

Application | Static electricity • Polarization

Theory | In this demonstration a $2' \times 4'$ wooden board is made to move by inducing an attractive force between a charged balloon and the wood. When the charged balloon is brought near the surface of one end of the board, the balloon induces electrons in the atoms at the wood's surface to move further into the balloon's surface, giving the surface a net positive charge. As a result, the board is attracted to the balloon. This process is known as charging by polarization. The induced charge is temporary as it is dependent upon the proximity of the charged balloon. Once removed, the polarized surface becomes neutral once again. Note that the polarized object remains electrically neutral—the electrons are simply redistributed, resulting in the objects developing positive and neutral poles. If balanced on a pivot, the board will rotate towards the charged balloon.

This demonstration is best performed in a dry environment (relative humidity $< 50\%$) since humidity causes static electric charge to dissipate more quickly. A dry board is preferred because the higher the moisture content in the wood, the greater its ability to absorb charge, since water is a highly polar molecule.

Materials | Adhesive, construction

Balloon, latex

Board, wooden, $2' \times 4'$, 1–2 m long

Watch glass, 10 cm

Wool cloth

Safety Precautions | Although this demonstration does not pose any obvious risks, always follow laboratory safety rules.

Preparation | Balance a $2' \times 4'$ wooden board on a watch glass to determine its balance point. Use a pencil to mark the location where the watch glass should be placed to balance the board. Use construction adhesive to permanently attach the watch glass to the board.

Demonstration | Balance the $2' \times 4'$ board and watch glass on the corner of a table or on top of an inverted beaker. Charge a latex balloon by rubbing it on your hair or with a wool cloth. Bring the charged balloon near one end of the balanced wooden board. The board should move toward the charged balloon. Discuss the term polarization and use it to explain why the board moved.

Generating Electrostatic "Phun"

Application | Static electricity • Van de Graaff generator • Electrostatic generators

Theory | Most Van de Graaff generators consist of a moving insulating belt that strips electrons from its base and transfers them to the dome of the device. The voltage on the dome can reach as high as 500,000 volts in a short period of time. The electric field of the generator radiates outward from the dome and any grounded object brought near the dome will serve as a conduit permitting the electrons to flow into the ground. The direction of the radiating field lines may be visualized by holding a candle near the dome—the flame will point away from the dome. Since the voltage is exceedingly high, even the air between the dome and a grounded object will become electrically conductive at distances as close as 20–30 cm, resulting in the electrons discharging as a sudden violent spark through the air. Insulated objects placed in contact with the generator will develop a charge approaching the charge of the generator. Objects such as a cluster of puffed rice, stacked pie plates and hair will separate from one another due to the like-charge.

Materials |

Candle, tapered, supported by a non-metallic candleholder

Donut-shaped dog's toy, rubber

Insulated platform such as a plastic step stool or plastic 4-gal milk crate

Matches

Paint stirrer, wooden or meter stick

Pie plates, aluminum (disposable), 5

Puffed rice, ¼ cup

Van de Graaff generator (or Wimshurst machine) and grounded wand with a ball

Safety Precautions | A typical Van de Graaff generator can produce voltages as high as 500,000 volts but with very low current. Despite the low current, certain precautions must always be taken. In particular, verify that volunteers who come near or come into contact with the generator do not have a history of heart trouble. When charging a volunteer in the "Fly-Away Hair" demonstration, be certain that the volunteer stands on an insulated platform and that his or her body is clear of any grounded surface, such as the edge of a table or a chair. Keep a wooden paint stirrer or meter stick on hand at all times to immediately turn off the generator in case of discomfort. Discharge by touching the dome with the grounded wand.

Preparation | Set up and test the Van de Graaff generator prior to performing any of the demonstrations. Follow the manufacturers' directions if adjustments are needed. Refer to the article listed in the reference section for maintenance and troubleshooting recommendations for working with Van de Graaff generators.

Demonstrations | **Sparky**

Turn on the generator and allow a charge to build up for about 30 seconds. Bring a grounded wand with a ball toward the dome of the generator until a spark jumps.

Review how the electric charge is produced by the generator, what type of charge is produced, and why the electrons jump between the dome and the grounded wand. Shut off the generator and discharge it by touching the dome with a grounded wand.

Blow out the Candle

Light the candle and bring it near the dome of a running generator until the flame is deflected away from the dome, being careful not to get close enough to cause a discharge. Move the candle around the dome to show that the flame is always deflected outward. Shut off the Van de Graaff generator and discharge it by touching the dome with a grounded wand.

Crackle and Pop the Rice

Place a rubber donut-shaped dog toy on top of the dome and pour about ¼ cup of puffed rice into the donut cavity. Turn on the Van de Graaff generator and watch as the rice "crackles" and "pops" out of the donut. Discuss how like charges cause repulsive forces to push the rice pieces away from each other and the dome. Repeat the demonstration and direct students' attention to the fact that the spray of rice pieces follows the field lines emitted from the dome of the generator. Shut off the Van de Graaff generator and discharge it by touching the dome with a grounded wand.

Flying Saucers

Place a stack of five inverted pie plates on top of the dome. Turn on the generator and watch as the pie plates fly off the stack one at a time. Ask students why the uppermost plate always flies off instead of two or more plates flying off together. Shut off the Van de Graaff generator and discharge it by touching the dome with a grounded wand.

Fly-Away Hair

Place the Van de Graaff generator near the edge of a table and position an electrically insulated platform about 50 cm away from the table. Select a volunteer with dry, untreated, long, thin, loose hair. Ask her to stand on the platform and place both hands firmly on the dome of the Van de Graaff generator. Instruct the volunteer not to remove her hands from the dome while the generator is running. Turn on the generator and verify that the volunteer does not experience any discomfort. (In case of discomfort, immediately then turn off the generator using a wooden paint stirrer to avoid unintentional discharge of the generator.) Ask the volunteer to shake her head from side to side until her hair fluffs outward and begins to stand up on end. End the demonstration by turning off the generator using the wooden paint stirrer, and then instruct the student to remove her hands from the generator. Direct students' attention to the fact that the volunteer's hair remains standing on end. At this point ask the volunteer to step down to the floor—this will allow the charge to discharge through the soles of her shoes, minimizing discomfort. Shut off the generator and discharge it by touching the dome with a grounded wand.

Reference Berg, Richard E. "Van de Graaff Generators: Theory, Maintenance, and Belt Fabrication"; *The Physics Teacher;* American Association of Physics Teachers, 1990, vol 28, p 281.

Keep It out of the Cage

A radio and cell phone are placed in various Faraday cages to illustrate how the electric field within the cage is zero.

Application

Faraday cage • Electric field • Electromagnetic waves

Theory

A Faraday cage is composed of an electrically conductive material surrounding a void. When an electromagnetic wave (including radio waves and microwaves) comes into contact with such an electrically conductive shell, the structure develops an electric charge. The charge is restricted to the surface since any charge attempting to penetrate the shell would immediately be repelled toward the outside. Consequently, the electric field inside the shell is always zero. In this demonstration an AM radio is placed inside a Faraday cage made of aluminum window screens. The radio will not operate when inside the cage since the radio waves are too long and consequently cannot penetrate the holes in the screen in order to create an electric field within the cage. A cell phone, however, will operate within the Faraday cage since the micrometer-sized waves are small enough. But, if the cell phone is wrapped in aluminum foil or placed inside a metal box, then the waves cannot pass between the atoms of the foil or metal. A car or airplane acts as a Faraday cage when struck by a bolt of lightning—the lightning bolt travels along the outer surface without doing damage to the inside.

Materials

Aluminum foil, $1' \times 2'$, or metal cookie tin	Hot glue gun
Cell phones, 2	Radio, small hand-held AM
Hot glue	Window screen, aluminum, $2' \times 2'$

Safety Precautions

Although this demonstration does not pose any obvious risks, always follow laboratory safety rules.

Preparation

Fold the aluminum window screen in half and create a pouch by sealing two edges using hot glue. This will be used to contain the radio or cell phone.

Demonstration

Tune a radio to a station with a clear signal. Place the radio inside the pouch made from the aluminum window screen. Direct students' attention to the fact that the radio stops receiving signals when it is inside the pouch. Discuss how the pouch serves as Faraday cage. Next, replace the radio with a cell phone. Use a second cell phone to call the phone inside the pouch. Ask students to propose a reason why one signal penetrated the cage while the other did not. Compare the frequency and wavelengths of the radio signal versus the cell phone signal.

Wrap a cell phone in aluminum foil or place it inside a metal cookie tin and once again call the phone. The phone will not ring this time, since the microwave signal cannot penetrate the foil. Unwrap the phone and call it again. Once the phone begins ringing, quickly wrap it in aluminum foil and then disconnect the calling phone. Direct students' attention to the fact that the circuitry of the receiving phone did not receive the hang-up signal and thus continues to ring.

A Leyden jar is used to demonstrate capacitance.

Application

Capacitance • Electric charge • Conduction

Theory

A Leyden jar is a device that can store electrical charge. It consists of a glass jar with an inner and outer coatings of thin metal leaf. A conducting rod, held to the jar with an insulating rubber stopper, is in contact with the inside of the jar. A Leyden jar is similar to a modern day capacitor.

A Leyden jar can be charged by transferring static electricity to it by conduction from another object. The greater the transfer, the greater the stored charge. The relative amount of stored charge can be shown using an electroscope.

As electric charge is added to the Leyden jar, the inner metal coating develops an overall negative charge. In turn, the outer coating of metal experiences a redistribution of negative charge. This results in some electrons moving away from the glass, creating an overall positive charge nearest the glass. As more charge is added to the inner layer, more electrons are displaced in the outer layer. The attractive forces between the positive layer in the outer metal film and the negative charges in the inner film are what keep the excess electrons inside the jar from escaping as more and more charge is added.

Materials

Electroscope

Fur or wool cloth

Leyden jar

Neon indicator lamp (e.g., RadioShack part #272-002)

Patchcord with alligator clips

Rod, rubber or plastic

Safety Precautions

Although this demonstration does not pose any obvious risks, always follow laboratory safety rules.

Demonstration

Connect a patchcord to an electroscope. Charge a rubber rod by rubbing it with a piece of fur or wool cloth. Touch the knob of the Leyden jar with the charged rod. Bring the free end of the patchcord that is connected to the electroscope near the knob of the Leyden jar. The leaves in the electroscope should separate to some degree. Charge the rubber rod and touch the Leyden jar again. When the electroscope is brought near the knob of the Leyden jar this time, the leaves should move even farther apart since the jar contains a greater electric charge.

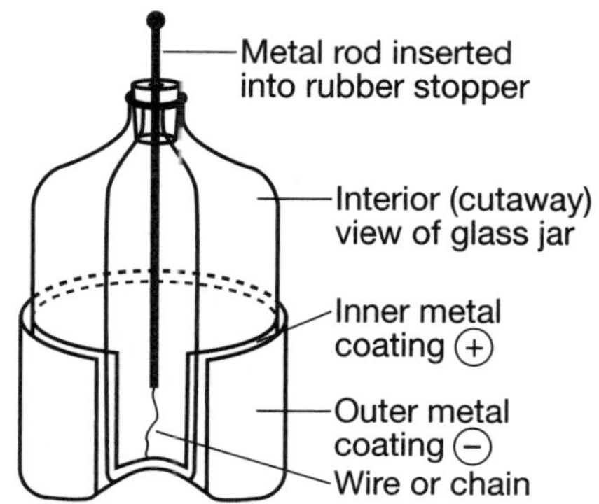

Electric Field Line Demonstrator

A homemade device is used to illustrate the nature of electric field lines.

Application

Electric field lines • Charge density

Theory

When a charged object is placed in an electric field, the object will follow a particular path due to the force of the field acting on the object. The path taken by the object is called a field line. If two or more objects enter the field, it becomes apparent that the field lines repel one another. In this demonstration a multi-surfaced object is covered with strands of Mylar tinsel (aluminum-coated plastic strips). When the object is charged, the strands of tinsel stand on end, indicating the location of the electric field lines. The lines do not intersect because of the repulsive forces between them. Furthermore, the lines radiate at right angles with respect to the tangent to the surface. This shows that the concentration of the electric field lines depends on the radius of the curvature of the object's surface. This concentration is called charge density. In this demonstration the lines are more concentrated on the tip of the cone because its radius is smaller than the opposite end.

Materials

Aluminum foil

Board, wooden, $5'' \times 5'' \times \frac{3}{4}''$

Christmas tinsel, thin ($<\frac{1}{8}''$ in width), $3''$ long

Dowel, wooden, $\frac{3}{8}'' \times 24''$

Drill

Drill bit, $\frac{3}{8}''$

Electrostatic generator (e.g., Van de Graaff generator)

Glue, cool melt

Glue, wood

Knife, large, bread

Styrofoam® ball, $4''$

Styrofoam® cone, $4''$ diameter $\times 10''$ tall

Tape, clear

Safety Precautions

A typical Van de Graaff generator can produce voltages as high as 500,000 volts but with very low current. Despite the low current, certain precautions should be taken. In particular, verify that volunteers who come near or come into contact with the generator do not have a history of heart trouble. When charging a volunteer as in the "Fly-Away Hair" demonstration, be certain that the volunteer stands on an insulated platform and that his or her body is clear of any grounded surface, such as the edge of a table or a chair. Keep a wooden paint stirrer or meter stick on hand at all times to immediately turn off the generator in case of discomfort. Discharge by touching the dome with the grounded wand.

Preparation

Use a bread knife to cut a Styrofoam ball in half. Use cool melt glue to attach one hemisphere to the bottom of a Styrofoam cone. Wrap the Styrofoam device with aluminum foil and use clear tape to secure the foil. The foil must be as smooth as possible since the static charge will quickly leak from all creases. Drill a $\frac{3}{8}''$ hole into the face of a small wooden board. Use wood glue to affix a wooden dowel in this hole. Allow the glue to set overnight. Pierce a hole into the side of the Styrofoam device and use glue to mount it horizontally on the protruding dowel. Use clear tape to cover the entire surface of the device with $3''$ pieces of tinsel evenly spaced about $\frac{1}{2}$–$1''$ apart.

Demonstration

Place the field line device about half a meter from the electrostatic generator. Activate the generator and watch the pieces of tinsel stand on end. Ask students if they can explain why the tinsel stands on end and why the individual pieces do not intersect. Point out that the strands of tinsel at the pointed tip of the cone appear to position themselves closer to one another than those at the rounded end. Discuss how electric field lines remain independent of one another and how the density of the electric field lines is dependent upon the shape and size of the surface.

Charge Density on a Balloon — *Size Matters!*

A charged balloon is used to illustrate how charge density depends on the radius of curvature of a surface.

Application	Charge density • Electric field lines
Theory	Charge density refers to the concentration of electric field lines. It depends upon the radius of curvature of a surface as well as the overall charge. For example, when two spheres of different sizes have the same charge, the smaller sphere will have the greater charge density.
Materials	Balloon, large Puffed rice cereal, ¼ cup Wool cloth
Safety Precautions	Although this demonstration does not pose any obvious risks, always follow laboratory safety rules.
Demonstration	Inflate a balloon but do not tie it off. Use a wool cloth to charge the surface of the balloon. Bring the charged balloon about an inch above some puffed rice piled on a table. Observe how the rice reacts to the balloon by leaping towards it and appearing to dance. Ask students to explain the reaction of the rice and discuss how the charged balloon induces a charge in the rice, causing the rice to move. Ask students what will happen if half of the air is released from the balloon and the demonstration is repeated without recharging the balloon. Release the air and repeat the demonstration. Observe how the rice dances with greater fury. Discuss how decreasing the radius of the balloon without changing its charge increases the charge density of the balloon.
Reference	Planinsic, Gorazd. "Increasing the Electric Field by Squeezing Charges"; *The Physics Teacher;* 2007; vol 45, p 393.

Chapter 14

Electric Potential and Capacitance

Air Capacitor

A mechanical analog is used to demonstrate the transfer of electric charge.

Applications | Capacitors

Theory | A capacitor is a device that stores electric charge. A typical capacitor consists of two parallel plates separated by a nonconductor. As the capacitor is charged, electrons collect on one side of the capacitor, creating a corresponding opposite charge on the other side of the capacitor. The opposite charge is induced by the creation of an electric field as the electrons enter the device. This demonstration uses a mechanical analog to represent this idea. The device consists of two air chambers separated by a rubber membrane. Air is introduced into one chamber, causing an increase in pressure, which in turn pushes the air out of the second chamber.

Materials | Adhesive, construction

Balloon, large or latex glove

Candle and holder

Matches

Nail, large

Pliers

Propane torch or Bunsen burner

Scissors

Tape, electrical

Tubing, ¼″ Tygon®, 30 cm (e.g., Aquarium pump tubing)

Tumblers, rigid plastic, 9-oz, 2

Safety Precautions | Propane is highly flammable and a fire risk. Wear safety glasses whenever working with flames, heat or glassware. Follow all laboratory safety rules.

Preparation | Using pliers, hold a large nail in the flame of a propane torch or Bunsen burner and heat the metal to redness. Pierce the bottom of each plastic tumbler (cup) using the hot nail and form a hole large enough to insert the ¼″ diameter tubing. Insert a 15 cm piece of Tygon tubing into each hole and make an airtight seal using construction adhesive. Allow the plastic cups to cool. Use a pair of scissors to cut the balloon or glove to form a circle large enough to cover the top of one of the cups. Stretch the rubber across the cup tightly and secure it to the cup using electrical tape. Cover this with the mouth of the second cup and tape the two cups together.

Review the components of a capacitor and how it works. Emphasize that the electrons entering one side of the capacitor do not travel out the other side, but rather push electrons out of the other side of the capacitor without crossing the barrier that separates the two sides. Show students the mechanical analog of a capacitor. Light the candle. Blow into the air capacitor with a quick puff of air while aiming the other end towards the candle flame. The flame should go out. Ask students why the flame went out. Emphasize that it is not the air from your lungs that extinguished the flame—the flame was extinguished by the air pushed out the other half (cup) of the capacitor.

Reference

Thanks to Elise Burns of Pascack Hills High School, Montvale, NJ for sharing this idea.

Cap Tester

A homemade device is used to illustrate the relative capacitance within a single capacitor.

Applications Capacitors • Capacitance • RC circuits

Theory This demonstration uses a 555 integrated circuit chip as a timing switch that sends a pulsating voltage to a speaker, thereby producing an audible tone. The pulse is controlled by a capacitor. The higher the capacitance, the lower the pitch of the tone. The cap tester, which you will create for this demonstration, can be used to compare one capacitor to another. The 555 chip charges the capacitor (a sawtooth wave) from pin 8 through R_1 and R_2 and then discharges to pin 7 through R_2, sending a pulsating voltage (a square wave) to a speaker. Electronic music synthesizers utilize similar circuits to produce musical tones. The tones are directly related to the capacitance in the circuit. The speaker may be replaced with a LED and used along with a 5μF versus 10μF capacitor to demonstrate how the turn signal in a car operates. Two different flash rates will be produced using these two capacitors. Other demonstrations using this cap tester are found later in this chapter.

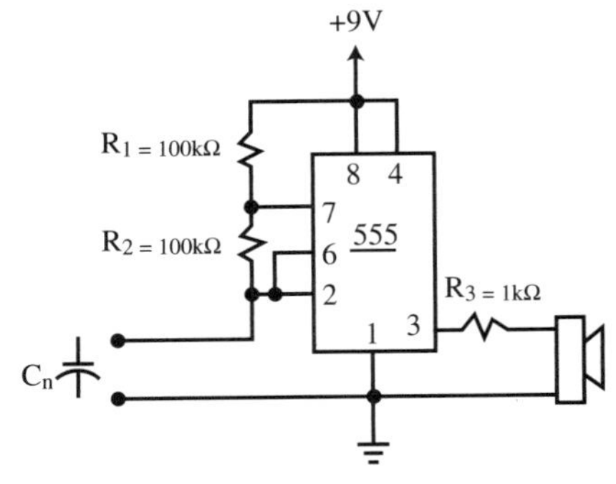

Materials Adhesive, construction

Alligator clips, 2

Battery clip/holder and battery, 9-V

Board, wooden, $4'' \times 4'' \times 1''$, for mounting circuit board

Capacitors (e.g., 0.01–100μF)

Piezo speaker (e.g., RadioShack 273-092, 8Ω, 0.1W)

IC socket, 8-pin

IC chip, 555

Project circuit board, $2'' \times 2''$

Resistors, 100 kΩ, 2

Resistors, 1 kΩ, 2

Soldering iron and solder

Wire, 22-gauge, $\approx$ 1 m

Wire strippers

Safety Precautions Although this demonstration does not pose any obvious risks, always follow laboratory safety rules. Follow manufacturers' directions when using soldering irons.

<table>
<tr><td>Preparation</td><td>Connect the circuit as illustrated in the diagram. The 9-volt battery is connected to pins 1 and 8. The piezo speaker is connected to pins 1 and 3. Cut two pieces of wire about 15 cm in length and use these to connect the alligator clips to pins 1 and 2. Use construction adhesive to mount the circuit board and battery holder to a wooden base.</td></tr>
</table>

Optional: The device may be modified to allow it to be suspended from a magnetic chalkboard or white board. To do this, use a ½″ Fosner bit to drill four holes into the base of the wooden board. Use construction adhesive to secure ½″ neodymium magnets in these holes.

Demonstration Connect the leads of the cap tester to different capacitors to relate the tone emitted by the tester to the capacitance of the device. The higher the pitch emitted by the tester, the lower the capacitance.

Pie Plate Capacitor

A homemade capacitor is used to demonstrate the variables that affect capacitance.

Application

Capacitance • Capacitor • Dielectric

Theory

A capacitor's ability to store charge depends on three factors: the distance that separates the plates (d), the cross-sectional area of the plates (A), and the dielectric properties of the material (K) in between the plates.

$$C = K\varepsilon_0 \frac{A}{d}$$

The constant ε_0 is the permittivity of free space, which has a value of 8.85×10^{-12} $C^2/N{\cdot}m^2$. Increasing the area of the plates provides greater room for electrons, which increases the stored charge. Decreasing the distance between the plates decreases the voltage between them. This results in an increase in the capacitance, C. The capacitance is also dependent upon the dielectric properties of the material separating the plates. A dielectric inhibits the transfer of charge across the plates. For example, wax paper is a better dielectric than copy paper. Capacitors are utilized in circuits to regulate current and voltage. Capacitors in photoflashes draw the energy stored in low voltage reservoirs (batteries). The capacitor has the ability to store higher voltages than the original reservoir. In turn, the photoflash capacitor is able to provide the high voltage required by the flash unit. Capacitors may also be used to block surges of voltage, which may damage an electrical device. Computer surge protectors are based on this principle. Capacitors can also limit the voltage emitted by a reservoir in the circuit.

Materials

Air variable capacitor (e.g., 384 pF air variable capacitor removed from an old AM radio)

Cap tester (see page 256)

Drill

Drill bits, $\frac{1}{16}''$ and $\frac{3}{16}''$

Dowels, wooden, $\frac{1}{2}'' \times 6''$, 2

Pie plates, aluminum (non-disposable), 2

Right angle clamps, 2

Ring stands, 2

Screwdriver

Screws, $\frac{1}{8}'' \times 1''$, 2

Tape

Wax paper, 25 cm

Safety Precautions

Although this demonstration does not pose any obvious risks, always follow laboratory safety rules. Follow manufacturers' directions whenever using power tools.

Preparation

Drill one $\frac{1}{16}''$ pilot hole into the center of one end of each dowel. Drill a $\frac{3}{16}''$ hole into the center of each pie plate. Use screws to connect the dowel to the inside of the pie plates. Mount the pie plates to separate ring stands using right angle clamps. Cover the face of one plate with wax paper and secure it with tape. Arrange the bottom of the plates so they face one another.

Demonstration

Direct students' attention to the parallel plates of the apparatus. Review the components of a capacitor. Emphasize that capacitance depends upon the area of the plates, the distance between the plates, and the dielectric material separating the plates. Connect the leads of the cap tester to the plates. Adjust the plates so that they directly face one another. Move the plates close together until a tone is heard from the cap tester. Move the plates even closer to observe how the pitch drops, indicating the relationship between capacitance and distance between the plates that is demonstrated by the tone. With the plates pressed against one another, slowly slide the plates off center in order to decrease the surface area between them. The tone will increase as the surface area decreases, indicating a decrease in capacitance. Realign the plates and press them together firmly. Have students make a mental note of the particular tone. Replace the wax paper with copy paper and compare the dielectric properties of the copy paper versus the wax paper. Finally, connect an air variable capacitor (the type used in AM radios) to the cap tester and show how the capacitance depends on the overlap of the plates.

Dissect a Capacitor

A commercial capacitor is dissected and another is constructed using everyday materials.

Application | Capacitors • Dielectrics

Theory | A capacitor consists of three main components: a pair of conducting plates, an insulating dielectric material separating the plates from one another, and a container to house and protect these pieces. Common dielectric materials include oil-soaked paper, plastics, glass, and minerals, such as mica.

Materials

Aluminum foil, $12'' \times 8''$, 2 pieces	Paper, copy, $8\frac{1}{2}'' \times 11''$, 3 pieces
Cap tester (see page 256)	Paper towel tube, trimmed to $10''$ length
Capacitor, cylindrical (e.g., 1-500µF)	Scissors
File folder	Wax paper, 25 cm
Gloves, latex	Wire cutters, sharp-tipped, or penny shears

Safety Precautions | Wear gloves when dissecting the capacitor to minimize exposure to the oil contained within the capacitor. Follow all laboratory safety guidelines.

Demonstration | Use a pair of sharp-tipped wire cutters or penny shears to cut open the container of the capacitor. Be careful so you do not damage the layers contained within. Gently pull out the rolled capacitor plates from the container. Slowly unwrap the plates to show that the capacitor is composed of four pieces: two conductive plates and two alternating insulating layers, one of which acts as the dielectric between the plates.

Using the dissected capacitor as a model, cut two $8'' \times 12''$ plates from aluminum foil. For each plate, trim a $1'' \times 7''$ piece off of one of the $8''$ sides leaving a $1''$ tab protruding from the aluminum plate (see diagram). Construct a capacitor by alternating the two pieces of aluminum foil plates between three sheets of paper. Arrange the aluminum tabs such that they protrude from opposite sides. Place the sandwich inside a file folder to keep the plates pressed together. Connect the capacitor to the cap tester and ask students to make a mental note of the tone they hear. The tone is directly related to the capacitance of the device. Remove the homemade capacitor, gently roll it into a cylindrical form, and slide it into a paper towel tube. Retest the rolled capacitor. The tone should be nearly the same, indicating that compacting the capacitor by rolling it into a small cylinder does not affect its capacitance.

Circuits of capacitors connected in a series versus a parallel arrangement are compared to one another.

Application | Capacitor circuits

Theory | When capacitors are combined, they may be arranged in the form of a chain called a series connection or in a parallel arrangement. When connected in parallel, the combination increases the capacitance because the electrons entering the circuit may distribute themselves over a larger surface area. A parallel combination of capacitors increases capacitance. The total capacitance is equal to the sum of the individual capacitance:

$$C_{total} = C_1 + C_2 + \dots C_n$$

In contrast, a series connection or combination of capacitors results in a decrease in capacitance because the capacitors must share the total voltage of the circuit:

$$V_{total} = V_1 + V_2 + \dots V_n$$

or $\dfrac{Q}{C_{total}} = \dfrac{Q}{C_1} + \dfrac{Q}{C_2} + \dots \dfrac{Q}{C_n}$ which is reduced to $\dfrac{1}{C_{total}} = \dfrac{1}{C_1} + \dfrac{1}{C_2} + \dots \dfrac{1}{C_n}$

Materials | Alligator patch cords

Board, wooden, $1'' \times 3'' \times 12''$

Cap tester (see page 256)

Capacitors, 6 (recommended: 5000 pF, 8200 pF, two 10^4 pF, and 10^5 pF)

Hammer

Label maker or permanent marker

Nails, $1''$ finishing, 12

Soldering iron and solder

Safety Precautions | Although this demonstration does not pose any obvious risks, always follow laboratory safety rules. Follow manufacturers' directions when using soldering irons.

Preparation | Arrange 10 nails in two rows arranged $2'' \times 2''$ apart down the length of the face of the board. Wrap the leads of the capacitors around pairs of nails beginning with the 5000 pF capacitor and ending with the 10^5 pF capacitor. Secure the connections by soldering the leads to the nails. Place prominent labels under each capacitor identifying their values. Test the board prior to using it in class.

Optional: The device may be modified to allow it to be suspended from a magnetic chalkboard or white board. To do this, use a $\frac{1}{2}''$ Fosner bit to drill four holes into the base of the wooden board. Use construction adhesive to secure $\frac{1}{2}''$ neodymium magnets in these holes.

Ask students if they can predict what will happen if two capacitors are connected. Discuss the physical difference in connecting the capacitors in series versus in a parallel connection. To familiarize students with the operation of the cap tester, connect the tester to the first three capacitors (having different values) one at a time. The tone emitted by the tester is directly related to the capacitance. The higher the pitch, the lower the capacitance.

Compare the tone emitted by the tester when it is connected to one capacitor versus when it is connected to two capacitors that are connected in parallel. Use alligator patch cords to make these connections. A lower tone is emitted when two capacitors are connected in parallel, indicating parallel connections increase capacitance. In contrast, when the cap tester is connected to two or more capacitors arranged in series, a higher tone is emitted, showing that the capacitance is decreased.

A charged capacitor is used to power a motor that lifts a weight.

Application | Capacitance • Energy • Electric potential • Voltage • Efficiency

Theory | In this demonstration a 1 F 5 VDC capacitor is charged using a 9-V battery. The capacitor is designed to be self-limiting and therefore will not charge beyond its 5 V limit unless the current is exceedingly high, which will cause the dielectric to break down. When the capacitor is charged, it can do work. That is, the stored energy may be released to move an object. The quantity of stored energy in a capacitor depends on its capacity to store charge (capacitance) and the potential difference (voltage) across the plates of the capacitor:

$$E = \tfrac{1}{2}\,CV^2$$

If the charged capacitor is connected to a motor, the capacitor will discharge, creating a current through the motor and causing it to lift the weight. In an ideal system, the change in gravitational potential energy of the weight is equal to the energy lost by the capacitor:

$$mgh = \tfrac{1}{2}\,CV^2$$

Most systems, including the one in this demonstration, are not ideal and subsequently some of the energy is lost and converted to heat. The most efficient motors barely approach an efficiency of 50 percent. In terms of work and energy, efficiency (%E) compares the work output to the work input (i.e., initial energy):

$$\%E = \left(\frac{\text{work output}}{\text{work input}} \right) \times 100\%$$

When a capacitor is charged, it has the potential to do work. That is, the capacitor contains stored energy—electric potential energy, EPE. When discussing the potential to do work (EPE), it is important not to confuse the term with electric potential. Electric potential is synonymous with voltage. Voltage is measured in Joules (energy) per unit charge (J/C), whereas electric potential energy is measured in Joules.

Materials |

Adhesive, construction	Mass, 100-g
Battery clip/holder and battery, 9 V	Multimeter
Capacitor, 5 VDC, 1 Farad	Right angle clamp
Dowel, 10″ × ½″	Ring stand
Double throw double pole (DTDP) knife switch	Soldering iron and solder
Epoxy, plumber's	Thread
Hand generator, Genecon™	Wire, 18–22 gauge, 1 m
Masonite board, 5″ × 7″ × 3⁄16″	Wooden dowel, ½″ × 6″

Note: The Genecon hand generator used in this demonstration includes a transmission gearbox to reduce the rate of lift, thereby reducing the strain on the motor. Alternate motor and gearbox systems, such as a Lego® motor and gearbox, may be used in place of the Genecon.

Capacitors are dangerous because they store electric energy that is easily released. Do not use capacitors rated higher than 5 VDC for this demonstration. Always follow all laboratory safety rules when performing demonstrations.

Preparation

Use construction adhesive to mount a 6″ × ½″ wooden dowel onto the nut securing the handle to the Genecon. Mount the body of the Genecon, the capacitor, switch, and battery clip to

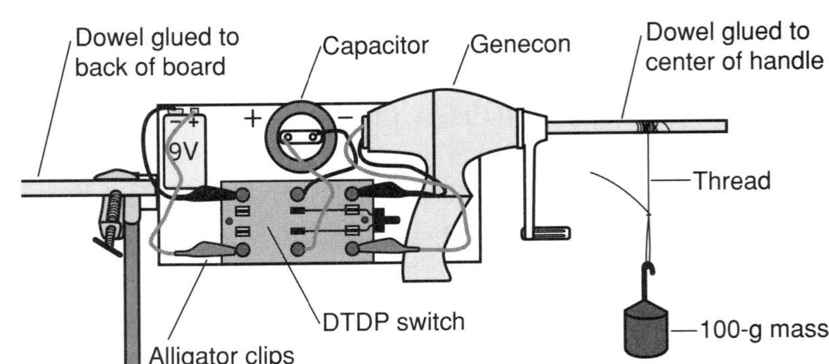

the face of the Masonite board using adehesive as illustrated in the diagram. Allow the adhesive to set overnight. Use a wire to connect the battery clip to the first pair of screws of the switch. The capacitor is then connected to the middle pair of screws, while the motor is connected to the third set. Be sure that the positive side of the capacitor is matched with the positive side of the battery. The leads to the capacitor and motor will require soldered connections. Use construction adhesive to attach a 10″ × ½″ dowel to the back of the Masonite—this dowel will be used to mount the apparatus to a ring stand. Tie a bowline knot (loop) on the end of a piece of thread and secure it to the dowel protruding from the handle using adhesive so the loop hangs 1 m below the motor. Use a right angle clamp to mount the apparatus to a ring stand. Suspend the weight from the loop. To prevent the battery from discharging during storage, position the switch to connect to the motor.

Demonstration

Discuss how the energy stored in a capacitor may be used to do mechanical work. Equate the energy stored in a capacitor to the work done in lifting an object. Review the components of the apparatus and how the three positions of the switch control the direction of the current. Connect a voltmeter to the center poles of the switch to which the capacitor is connected and verify that the capacitor has no electrical potential (i.e., voltage). Since the voltage is zero, the capacitor does not have any stored energy. This is an opportune time to discuss the relationship between stored energy, capacitance and electric potential ($E = ½\,CV^2$). Throw the switch toward the battery to charge up the capacitor. Note that the capacitor only charges to 5 V—not as high as the 9 V available from the battery. Return the switch to the center position and point out that the capacitor holds the charge over time. Finally, move the switch toward the motor position and observe how the motor lifts the weight. Measure the height that the weight is lifted and compare the change in gravitational potential energy to the initial energy in the charged capacitor. Compare the actual gravitational potential energy gain of the weight to the energy released by the capacitor and calculate the efficiency of this motor.

A capacitor is charged and discharged through a lightbulb to observe its operation.

Application

Capacitance • Time constant

Theory

In this demonstration a capacitor is charged in series with a lightbulb. Initially, the bulb glows brightly due the maximum current flow, but as the capacitor charges, the brightness and current decrease over time. Once the capacitor is fully charged, the current stops flowing. The battery is removed from the circuit and the bulb is connected only to the capacitor. Initially, the bulb glows brightly, but as the capacitor discharges, the brightness decreases over time. This demonstration shows that capacitors require time to charge and discharge. This property is referred to as the time constant of the capacitor. Most time constants follow an exponential voltage increase/decrease. The cap tester used in previous demonstrations depends upon this principle. The pulse rate of the tone is directly related to the time constant of the capacitor added to the circuit.

Materials

Adhesive, construction

Battery holder, $2 \times AA$

Bulb socket

Capacitor, electrolytic, 1 Farad, 5-V

Flashlight bulb, 3-V

Masonite board, $\frac{3}{16}'' \times 4'' \times 6''$

Screwdriver, flat head

Soldering iron and solder

Single pole double throw (SPST) knife switch

Wire, 22 gauge 15 cm long, 3 pieces

Wire strippers

Safety Precautions

Capacitors are dangerous because they store electric energy that is easily released. Do not use capacitors rated higher than 5 VDC for this demonstration. Always follow laboratory safety rules when performing demonstrations.

Preparation

Use construction adhesive to mount the capacitor, bulb socket, battery holder, and switch to the face of a Masonite board. Allow the adhesive to set overnight. Use a screwdriver and/or solder to connect the components and wire as illustrated in the diagram. Store the apparatus with the switch connected to the B-position to avoid discharging the battery.

Sketch a diagram of the circuit on the marker board. Show students the circuit and throw the switch to the A-position. Direct students' attention to the brightness of the bulb and how it changes. Ask students to explain why the bulb goes out. Following discussion, ask students what will happen if the switch is thrown into the B-position. Direct students' attention to the brightness of the bulb and how it changes. Discuss how capacitors work and what they are used for.

Reference

Bilash, Borislaw. *A Demo A Day—A Year of Physical Science Demonstrations.* Flinn Scientific: Batavia, IL, 1997; p 277.

Chapter 15

DC Circuits

Burn a Resistor

The current through a resistor is measured as the voltage is changed to show the relationship summarized by Ohm's Law.

Application Resistance • Current • Ohm's Law • Resistors • Power

Theory Ohm's Law summarizes the relationship between voltage, current, and resistance: $V = IR$. In this demonstration a resistor is connected to a power supply. The voltage is raised incrementally and the current is measured using an ammeter. Students will notice that the ratio between the voltage and current is constant for any voltage–current pairs. In terms of Ohm's Law, this constant is the value of the resistor. If the current exceeds the limitations of the resistor, the crystal structure of the resistor will break down, causing it to lose its ability to conduct current.

Materials Adhesive, construction

Binding posts, standard, 2 (available from electronic supply stores)

DC power supply, 0–5 volts, 1 amp minimum

Dowel, wooden, 6″ × ½″

Drill

Drill bit, ¼″

Multimeters with alligator patch cords, 2

Masonite board, 4″ × 4″ × 3⁄16″

Resistor, ceramic, 10-ohm, ⅛-watt

Right angle clamp

Ring stand

Optional: If a DC power supply is not available, then substitute eight 1.5-V batteries connected in series with a 100-ohm potentiometer.

Safety Precautions Follow manufacturers' instructions whenever using power tools. Keep a fire extinguisher at hand to extinguish flames. Keep flammable clothing and other materials away from the resistor. Always follow all laboratory safety rules.

Preparation Drill two ¼″ holes one inch apart along one edge of the Masonite board. Mount the binding posts in these holes. Use construction adhesive to attach a wooden dowel to the back of the board so it protrudes like a handle. Allow the glue to set overnight.

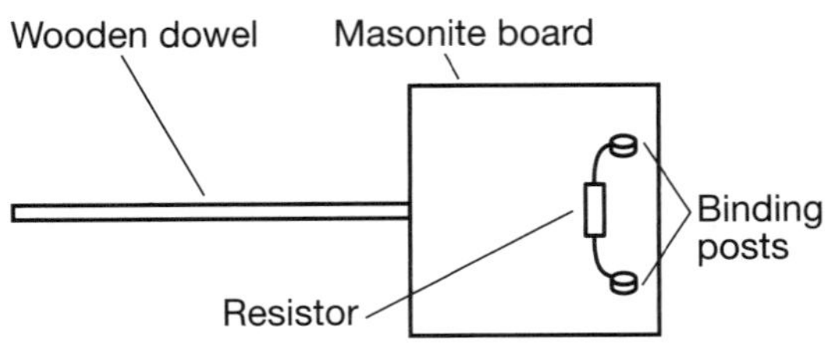

Mount the board on a ring stand using a right angle clamp. Connect a 10-ohm resistor across the binding posts. Attach one multimeter as a voltmeter in parallel with the resistor. A second multimeter will serve as an ammeter and is attached in series with the power supply and the resistor.

Demonstration

Discuss the relationship between voltage, current, and resistance in a simple DC circuit. Introduce the apparatus and explain that the circuit consists of a simple loop. Beginning with the voltage equal to zero, slowly increase the voltage to about half a volt. Have students write down the current reading at this particular voltage. Make four or five additional voltage/current pair readings, each time increasing the voltage about half a volt. Ask students to reflect on the voltage/current pair readings and see if they can discover a pattern. No doubt some students will quickly realize that the voltage/current ratio produces a constant number. In terms of Ohm's Law, this constant is the value of the resistor. Finally, increase the voltage slowly until the resistor begins to emit smoke. Continue to increase the voltage until the resistor bursts into flames. Bring students' attention to the fact that the current reading is now zero, but the voltage does not drop. Discuss why the current is now zero and why the voltmeter does not read zero.

Optional: This demonstration may be revisited once students have learned about electrical power. As a power demonstration, have students make a note of the voltage and current reading at the point when the resistor begins to smoke. Use these readings to calculate the power rating of the resistor using: $P = IV$.

Stopper My Flow

An analogy to illustrate how resistors limit current flow.

Application	Resistance • Current • Resistivity • Electromotive force
Theory	A resistor is a device that limits the current flow in a circuit. The degree of resistance is dependent upon the resistivity and the length and cross-sectional area of the material from which the resistor is made. Resistivity is specific to a material and dictates how well current flows through it.

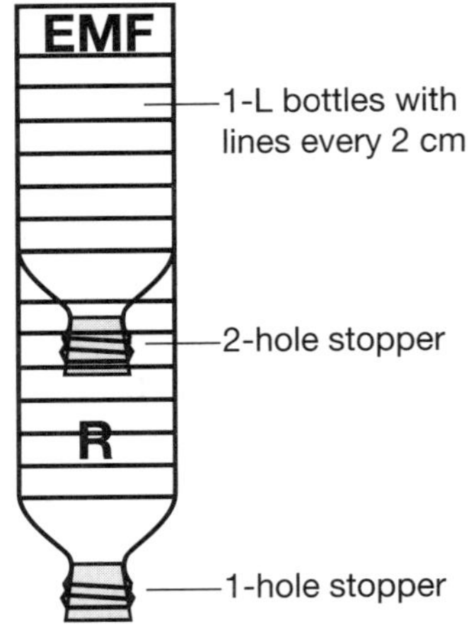

Materials

Beaker, 2-L or plastic pitcher

Marker, permanent, fine-tipped, black

Razor knife

Rubber stopper, 1-hole, #4

Rubber stopper, 2-hole, #4

Sink or bucket

Soda bottles, plastic, 1-L, 2

Water, colored red

Safety Precautions

Although this demonstration does not pose any obvious risks, always follow laboratory safety rules. Clean up water spills immediately.

Preparation

Use a razor knife to trim off the bottom of each soda bottle (about 2″). Remove the labels and use a permanent marker to draw 6–7 lines that encircle the body of each bottle every 2 cm. Use the marker to label one bottle "EMF" and the other "R." Place a 2-hole #4 rubber stopper into the spout of the EMF-labeled bottle and a 1-hole #4 stopper into the R-labeled bottle. The bottles must fit tightly together. No air or very little air should leak from the inner chamber, otherwise the flow from the upper 2-hole stopper will overtake the flow from the lower 1-hole stopper and the analogy will fail. Clear tape may be wrapped around the connection between the bottles to ensure an airtight fit.

Demonstration

Hold the bottle system over a sink or bucket. Fill the EMF bottle with colored water to begin the "current flow." Direct students' attention to the level of the water in the R bottle (the resistor model) as the "current" flows from the EMF. The level should remain constant, showing that the resistor bottle limits the "current" flow. Discuss how this model serves as an analogy of a resistor and how it limits the current flow in a circuit. Discuss how the number of holes in each stopper is analogous to the resistivity of the stopper.

Soda Bottle Potential

An analogy to illustrate the potential energy of a circuit.

Application | Potential difference • Electromotive force • Voltage • Battery

Theory | A DC power source (a battery or DC power supply) provides an electrical potential difference between one side of a circuit and the other. The greater the potential difference between the two sides of the power source, the greater the current flow through the circuit (with the same components). Electric potential difference is analogous to the gravitational potential energy of a waterfall. The higher the top of waterfall, the greater its gravitational potential energy. An electric potential difference exists between the two terminals of a battery. The magnitude of the potential difference in a battery is dependent upon its chemical composition. Common batteries are designed to have a potential difference of 1.5 volts. The potential difference of a battery-powered circuit may be increased by connecting two or more batteries in series. Potential difference is also known as voltage or electromotive force (EMF).

In this demonstration two bottles are used to represent the two sides of a battery. The "V+" side represents the high potential side of the battery (the positive terminal), whereas the "V–" side represents the low potential side (the negative terminal). The tubing represents the components of the circuit upon which the potential will act (e.g., wire and a lightbulb). To simulate the potential difference, the "V+" bottle is raised above the "V–" bottle, resulting in the flow of "current." Raising the "V+" bottle even higher raises the potential difference, causing an increase in current flow—analogous to adding an additional battery to the circuit.

Materials | Marker, permanent

Nail, large

Pliers

Propane torch or Bunsen burner

Sealant, silicon-based (for an aquarium)

Soda bottles, 1-L, not-colored, 2

Stopwatch

Tubing, aquarium, ¼″, 1 m

Water, colored red

Optional: A mechanical flow rate meter may be placed in line between the bottles to simulate the work done in the circuit.

Safety Precautions | Wear gloves and safety glasses when working with open flames. Follow laboratory safety rules.

<table>
<tr><td>Preparation</td><td>

Remove the labels from two 1-L soda bottles and use a permanent marker to draw a bold line that encircles the body of each bottle about ¾ length from the bottom. Mark one bottle "V+" and the other bottle "V–" using large lettering. Holding a nail with the pliers, use a propane torch to heat the nail red hot and then pierce a hole in the side wall of each bottle about 3 cm from the bottom. The hole should be just large enough for the aquarium tubing to fit snugly. Insert the tubing into the holes so the tubing connects the two bottles. Seal and secure the tubing to the bottles using silicon-based aquarium sealant. Allow the sealant to set overnight. Tighten the cap on the "V–" bottle and fill the other to the inscribed line with red-colored water. The cap will prevent the water from draining from one bottle to the other.

</td></tr>
<tr><td>Demonstration</td><td>

Discuss how our understanding of gravitational potential energy may be used as an analogy to understand electric potential energy. Raise the high-potential bottle (V+) just above the low-potential bottle (V–). Explain how this system represents the two sides of a battery (or DC power supply), with the tubing representing the components of the circuit. "Flip the switch" by removing the cap from the low potential end. The water is expected to flow from "V+" to "V–." Ask students to explain why the water flows in terms of gravitational potential energy. Use this analogy to discuss why current also flows from a high to a low potential. Refill the bottles to their initial conditions and repeat the demonstration, but this time measure the time it takes for the water to flow from the high-potential bottle and fill the low-potential bottle. Finally, repeat the demonstration a third time—this time raising the "V+" bottle to its highest potential. Discuss how the flow rate (current) increases when the potential difference is increased.

</td></tr>
</table>

A DEMO A DAY

A circuit of two bulbs is used to demonstrate the relationship between luminosity, resistance and temperature.

Applications | Resistance • Luminosity

Theory | An incandescent lightbulb consists of a coil of wire, called a filament, housed inside a glass enclosure. This enclosure is filled with an inert gas under a partial vacuum to prevent the filament from burning when it is lit. Lightbulbs are rated in terms of wattage or power consumption. The higher the wattage, the brighter the bulb. The brightness is due to the fact that the higher wattage bulb has a lower resistance. Consequently, the filament permits a higher current flow. In this demonstration two lightbulbs are connected in series—one with a low wattage (2.8 watts) and a second with a much higher wattage (25–100 watts). The higher wattage bulb serves as a resistor in this circuit, allowing the relationship between luminosity, current, and wattage to be studied. The high-wattage bulb is then replaced with a bulb from which the glass housing has been removed. The voltage in the circuit is low enough not to cause the filament to burn. When air is blown across the filament, the low-wattage bulb glows brighter, indicating an increase of current in the circuit. This shows the fact that resistance is temperature dependent.

Materials | Adhesive, construction

Battery clip/holder, 9-V

Battery, 9-V

Bell wire

Board, wooden, $1'' \times 5'' \times 12''$

Hammer

Lightbulb socket, standard

Lightbulbs, standard, 25 watt, 2

Lightbulb, standard, 60 watt

Lightbulb, standard, 100 watt

Lightbulb socket, mini screw

Lightbulb, type 1487, 14-V, 0.20 amps

Pliers, needle-nose

Single pole single throw (SPST) knife switches, 2

Safety Precautions | Wear safety glasses whenever working with flames, heat or glassware. Dispose of broken glass carefully. Always follow laboratory safety rules.

Preparation   Mount the two light sockets on either end of the wooden board using construction adhesive. Mount the 9-V battery clip and switches in between the two sockets. Use a bell wire to link the sockets and one switch ("switch 2") in series with the battery clip. Attach the other switch ("switch 1") in parallel with the large socket. Wrap one 25-watt bulb in a few layers of paper towels, place it on a hard surface and tap the bulb with a hammer to break the glass. Use a pair of needle nose pliers to remove any remaining pieces of glass attached to the bulb. Be careful not to break the filament of the bulb. Throw away the glass shards in a sharp objects receptacle.

Demonstration   Insert a 100-watt bulb into the apparatus. Show students the device and direct their attention to the fact that the two bulbs are in series with the battery. Close switch 1 and then switch 2. Ask students what will happen if switch 1 is opened. After discussing the various ideas, open switch 1. The small lightbulb should glow dimmer. Ask students what will happen if the bulb is replaced with another with different wattage. Observe what happens when the 100-watt bulb is replaced with a 60-watt bulb and then with a 25-watt bulb. Discuss the relationship between a bulb's wattage and its resistance in a circuit.

Replace the 25-watt bulb with the modified 25-watt lightbulb without the glass housing. Blow gently on the filament of the bulb and watch as the smaller bulb glows brighter. Ask students to propose an explanation.

Optional: The filament may be immersed in liquid nitrogen to show a dramatic increase in brightness due to the extreme decrease in resistance due to cooling.

Reference   John Valente, Marine Academy of Science and Technology, Sandy Hook, NJ, 2007.

A DEMO A DAY

A nickel–chromium wire inserted into a circuit is used to show the relationship between its length and the resistance it provides.

Application

Resistance • Resistivity

Theory

A resistor is a device that limits the current flow in a circuit. The degree of resistance is dependent upon the resistivity and the length and cross-sectional area of the material from which the resistor is made. Resistivity is specific to a material and dictates how well current flows through it. Some light-dimmer switches and audio-volume rheostats use this principle by varying the length of the contact position inside the switch. The nickel–chromium wire used in this demonstration has a gauge of 24–30 and a resistance of 5.48–22.14 ohms/m.

Materials

Adhesive, construction

Alligator clip

Battery holder, C or D cell with batteries, 2, 3 or 4

Bell wire, 1 m

Board, wooden, $4'' \times 1'' \times 3'$

Lightbulb socket, mini screw

Lightbulb, mini, 3–6 V

Pliers

Screw eyes or binding posts, $\frac{1}{2}''$, 2

Voltmeter (optional)

Wire strippers

Wire, nickel–chromium, 24–30 gauge, or guitar string, 1 m

Safety Precautions

Although this demonstration does not pose any obvious risks, always follow laboratory safety rules.

Preparation

Insert screw eyes (or binding posts) into the lower-opposite corners of the wooden board. Label one screw eye "A" and the other one "B." Use pliers to secure the Ni–Cr wire tightly from one screw eye to the other. Use construction adhesive to mount a lightbulb socket and a battery holder to the face of the board. Use bell wire to form a series circuit with the battery pack, lightbulb, and Ni–Cr wire, leaving one end open. Attach an alligator clip to the open end.

Demonstration

Connect the alligator clip to the screw eye labeled "A." In this position the Ni–Cr wire is bypassed. Ask students what will happen if the Ni–Cr wire is included in circuit. Then slide the alligator clip along the Ni–Cr wire to show how its length varies the resistance as is demonstrated by the change in light intensity. A voltmeter connected in series with the circuit may be used to measure the voltage as the length is varied.

Series versus Parallel Circuits

Series and parallel circuits are compared.

Application | Electric current • Series circuit • Parallel circuit

Theory | An electric circuit provides a complete and closed path for an electric current. An electric circuit consists of a load or resistance, wires, switch, and a source of electrons (power source). If all the parts are connected one after another, the circuit is said to be in series. In a series circuit, there is only one path for the electrons to take. If any part of the circuit breaks, the entire circuit is open and no current flows. Inexpensive Christmas lights are usually connected in series, because less wire is needed.

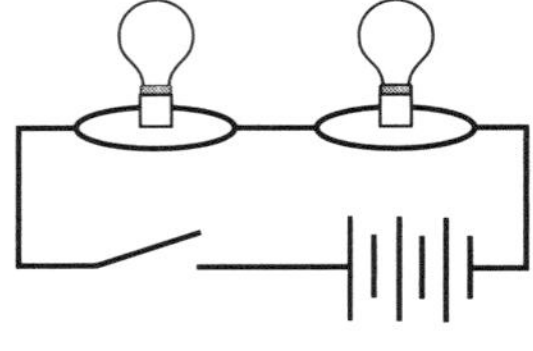

In a parallel circuit, different loads are on separate branches of the circuit. There are several paths for electrons to take in a parallel circuit. If one branch breaks, then the electrons flow through the remaining branches. Circuits in home wiring are parallel so that the entire circuit does not go out when a single lightbulb burns out.

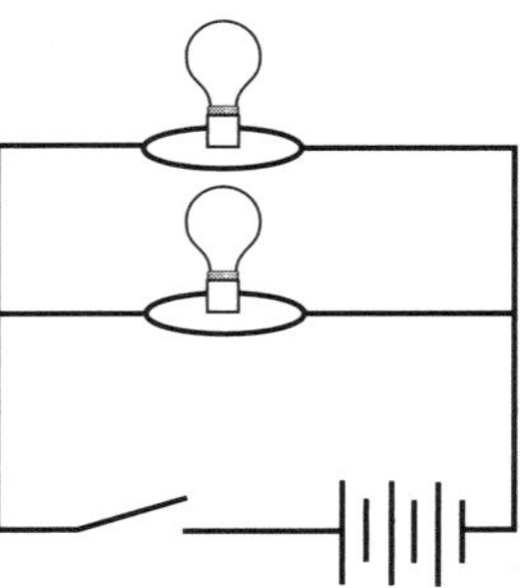

This demonstration uses a special apparatus that consists of two electrical outlets— one wired parallel and the other in series. Lightbulbs added to the parallel outlet light up as expected. Bulbs connected in series will glow dimly since each bulb receives less than its proper voltage (i.e., there is a drop in voltage as the current passes through each lightbulb).

Materials | 2-pole, 3-wire outlet, black or brown

2-pole, 3-wire outlet, white

Double outlet box, plastic

Double outlet cover plate

Electrical power cord with 3-prong plug

Insulated wire, 14 gauge, 50 cm

Lightbulb screw-plug-in adapters, 2

Lightbulbs, 15-watt, 2

Screwdriver

Wire strippers

Safety Precautions | Be sure to ground the device as indicated in the diagram. Do not use this device as a power strip or extension cord. Use for demonstration purposes only! Do not allow students to handle the device. Follow all laboratory safety rules.

Remove the metal strip joining the two brass colored screws on the white outlet, as illustrated in the diagram. The other pair of screws remain joined by a similar metal link. This outlet will serve as the Series Outlet.

Connect the two outlets and power cord as illustrated in the diagram. Connect the green (ground) wire to the round prong of the plug, the white (neutral) wire to the wide flat prong and the black (hot) wire to the narrow flat prong. Secure the outlets in a plastic electrical box and cover with a double outlet cover plate. Be sure that no wires are exposed.

Prepare the light sources by screwing two 15-watt lightbulbs into plug-in adapters.

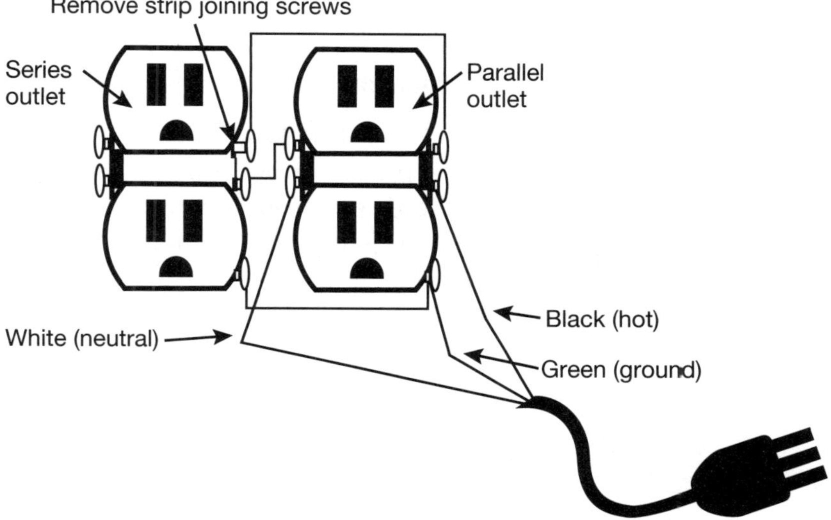

Demonstration

Plug in the series–parallel demonstration device. Place the lightbulbs into the parallel outlet. Both lightbulbs will glow brightly. Discuss how the current is shared in a parallel circuit. Remove one bulb to reinforce the concept that the current still has a pathway regardless of the number of bulbs connected in parallel.

Unplug the device and then transfer both bulbs to the other outlet—without revealing that this outlet is connected in series. Plug in the device. Both bulbs will glow dimmer in relation to the first outlet. Ask students to explain why the bulbs are dimmer this time. Discuss how the current must travel through both bulbs in series and consequently the voltage is shared between the bulbs. Finally, remove one bulb resulting in the other bulb going out to demonstrate that the current must pass through both bulbs to complete the circuit.

Reference

Harry Rheam, Physics 3 Team of the Woodrow Wilson Institute, 1991.

Kirchhoff Rules!

Kirchhoff's Laws are illustrated using two different circuits.

Application | Kirchhoff's Laws • Series circuit • Parallel circuit

Theory | The sum of the voltage in a series circuit must add up to zero. This is known as Kirchhoff's Voltage Law. A circuit consists of a source of voltage (a battery) and a resistance or load (lightbulb or motor). The current in the circuit is dictated by Ohm's Law: $I = V/R$. If an additional resistance, such as a second lightbulb, is introduced into this circuit, then the total current in the circuit will drop, but the current through each resistance will be identical. The sum of the voltages across each of these resistors will equal the voltage of the power source. In this demonstration lightbulbs of different ratings are utilized in the series circuit and consequently vary in brightness. This illustrates the contrasting voltage across each lightbulb.

Kirchhoff's Current Law states that if a current encounters a junction in a circuit, then it will split. One part will flow into one leg with the remaining current flowing into the other leg. The distribution of the current depends on the resistance in each leg. A greater resistance permits less current to flow. If the resistance is equal, as is initially the case in this demonstration, then the current is split evenly. The brightness of each bulb will be the same, as the brightness is directly related to the voltage across the bulbs. However, if bulbs of different ratings are inserted into the parallel circuit, they will still emit the same brightness. Although the current is dependent upon the resistance of the individual bulbs, the voltage drop across each bulb must be the same, according the Kirchhoff's Voltage Law.

Materials | Adhesive, construction

Alligator clip

Battery holder, C or D cell with batteries, 4×1.5-V

Battery and holder/clip, 9-V

Bell wire, 3 m

Boards, wooden, $5'' \times 1'' \times 12''$, 2

Lightbulb sockets, mini screw, 6

Lightbulbs, mini, of identical ratings (e.g., 6.15V/0.5), 3

Lightbulbs, mini, of different ratings (e.g., 1.2V/0.22A, 2.47V/0.3A and 6.15V/0.5A)

Single pole single throw (SPST) knife or wall light switches, 4

Voltmeter

Wire strippers

Safety Precautions

Although this demonstration does not pose any obvious risks, always follow laboratory safety rules.

Preparation

To prepare the series board, use construction adhesive to mount three light sockets, one knife switch, and a 9-V battery holder/clip to the face of a wooden board. Use bell wire to connect these components in series. To prepare the parallel board, use construction adhesive to mount three light sockets, three knife switches, and a C or D cell battery holder to the face of a wooden board. Use bell wire to connect these components in parallel, as illustrated in the diagram. Connect the three bulbs with different ratings to the series board and three lightbulbs with identical ratings to the parallel board.

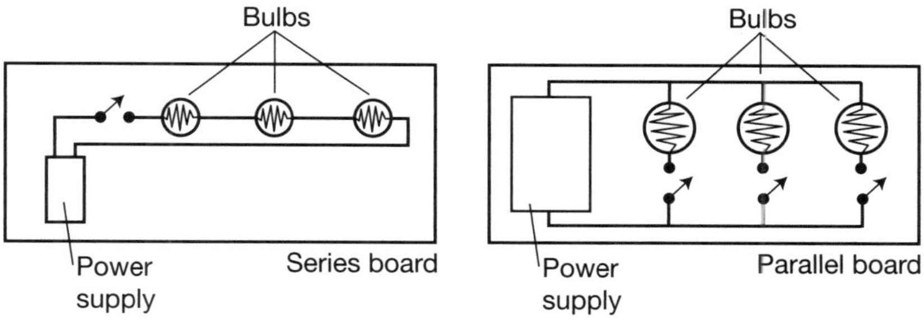

Demonstration

Turn on the series board and direct students' attention to the contrasting brightness of the three lightbulbs. Ask students to propose an explanation as to why the three bulbs vary in brightness despite being connected to the same source of voltage. Measure the voltage across each individual light socket and compare the combined voltage of the lightbulbs to the voltage across the battery. Discuss Kirchhoff's Voltage Law.

Next, turn on the lightbulb nearest the battery on the parallel board. Ask students what will happen to the brightness of the first lightbulb if an identical lightbulb is added to the circuit in parallel. Following discussion, turn on the second lightbulb and then the third. Ask students to explain how a parallel circuit differs from a series circuit in terms of current and voltage. Measure the voltage across each individual light socket and compare these voltages to the voltage across the battery. Discuss Kirchhoff's Current Law.

Finally, ask students to predict what they will observe if the identical bulbs in the parallel board are replaced with bulbs of different ratings prior to exchanging the lightbulbs. Remove one of the batteries from the holder, otherwise the power source will burn out the two smaller lightbulbs. Replace the bulbs and the battery. Discuss the results.

Light Puzzle

Students analyze a combination of lightbulbs connected to a hidden circuit applying their understanding of circuits and Kirchhoff's Laws.

Application Kirchhoff's Laws • Parallel circuits • Series circuits • Ohm's Law

Theory This demonstration uses a puzzle consisting of three lightbulbs connected to three switches and a power supply. Although the lightbulbs are visible, the connections between the components are hidden. Students observe whether or not the bulbs light up and their relative brightness as the switches are manipulated to determine the schematic of the circuit. The switches change the circuit from series to parallel to a combination of series and parallel circuits with each new permutation. With three switches eight combinations of lights and brightness are possible:

Permutations of Circuit 1

Permutation	Switch 1	Switch 2	Switch 3	Light A	Light B	Light C
1	Off	Off	Off	Off	Off	Off
2	On	Off	Off	On	On	On
3	Off	On	Off	Off	Off	Off
4	Off	Off	On	Off	Off	Off
5	On	On	On	Max	Max	Max
6	Off	On	On	Off	Off	Off
7	On	On	Off	Off	Off	On
8	On	Off	On	On	Off	Off

To make the puzzle more challenging, mount Light A over Switch 2. You may also choose to add additional lightbulb and switch pairs to the schematic

Materials Adhesive, construction

Batteries, AA, 2

Battery holder, AA, 2

Bell wire, 18–22 gauge

Cover plate, 3-switch, white plastic

Drill

Drill bit, $\frac{1}{8}''$

Light sockets, mini, 3

Light switches, household, 3

Lightbulbs, mini flashlight, 2.47V/0.3A, 3

Screwdriver

Triple gang-box

Wire cutters

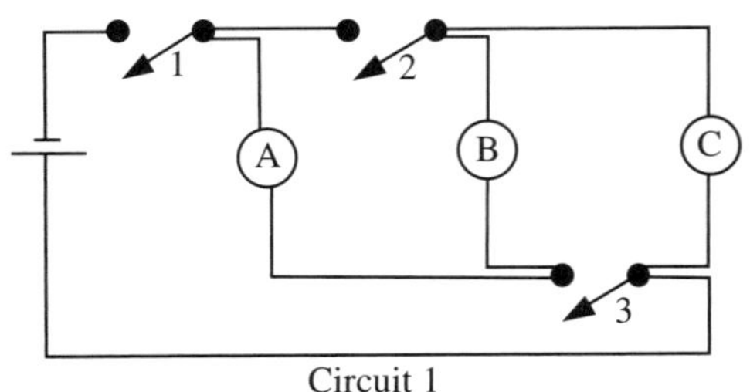

Safety Precautions	Although this demonstration does not pose any obvious risks, always follow laboratory safety rules. Follow manufacturers' directions when using power tools.
Preparation	Mount one light socket directly above each switch on the top of the gang-box using construction adhesive. Mount the battery holder on the side of the gang-box using adhesive. Allow the glue to set overnight. Drill a pair of $\frac{1}{8}''$ holes on either side of the light sockets. Also drill a pair of holes for the leads of the battery holder. Connect the battery holder, light sockets and light switches according the schematic. Attach the light switch cover onto the gang-box. Insert the batteries and test the device. Mini Christmas lights may be used in place of the mini flashlight bulbs.
Demonstration	Show students the light puzzle and assign them the task to uncover how the light-bulbs, switches and power supply are linked. Turn each switch on one, by one, asking students to decide which switch to begin with. Whether or not a light comes on and its brightness will depend upon which combination of switches is turned on. After trying all eight combinations ask the students to sketch the circuit.
Reference	Brian Holton, Rutgers University; New Brunswick, NJ, 1993.

Pop-It!

A number of appliances are connected to the same circuit to determine the amount of load needed to overload the circuit.

Application | Power • Current • Voltage • Heat • Expansion

Theory

The power in a circuit is equal to the product of the current and voltage in the circuit: $P = IV$. In this demonstration a number of appliances are connected to a circuit until the breaker is overloaded. Both the current and the voltage of the circuit are measured to show that only the current increases as the appliances are added. This is because household circuits are wired in parallel. As the current increases, it is accompanied by a rise in temperature of the wires. To prevent a fire, household circuits contain safety devices called breakers (or fuses) that interrupt the current in the event that the temperature reaches a set limit. Some breakers contain a bimetallic strip that, when heated, bends away from a connection, causing a break in the current flow. Most wall outlets are limited to 15 or 20 amps. A 1500 W hair dryer draws about 12.5 amps, leaving 2.5 amps for other appliances. A 15-amp circuit can handle thirty 60 W lightbulbs. During the summer months, a power company must often reduce the voltage to a community (called a "brownout") because of the high demand for power. This increases the likelihood of popping a breaker. The breakers are more apt to pop because the decrease in voltage is accompanied by an increase in current. If decreased to 110V, a 15-amp circuit can handle only twenty-five 60 W lightbulbs. Contrary to popular belief, the lights in your home are not dimmer during a brownout. If the voltage is suddenly reduced from 120V to 110V, the lights will momentarily dim but the subsequent increase in current will return the lights to their normal intensity. Connecting extension cords end to end is not advised because this will result in an increase in electrical resistance that lowers the voltage to a device. Doing so puts the device at risk of "burning out" because the increase in resistance is accompanied by an increase in current: $P = IR^2$.

Materials

Appliances, various (e.g., hair dryer, vacuum, clock radio, fan, toaster, blender, etc.)

Banana plug sockets, 2

Drill

Drill bit, ¼″

Multimeter with banana connectors

Multimeter with probe connectors

Outlet box, single

Outlet cover plate, dual

Outlet, single, female, with switch

Plug, 3-prong and 4′ wire assembly

Power strip with 15-amp breaker

Soldering iron and solder

Wire strippers

Safety Precautions

Follow manufacturers' directions whenever using power tools and soldering irons. Test the completed circuit using a multimeter and verify that the circuit is correctly polarized. Always follow all laboratory safety rules when performing demonstrations.

Drill a pair of ¼″ holes into the left side of the outlet box about 1–2″ apart. Insert the banana plug sockets in these holes. Trim a 6″ piece of wire from the plug–wire assembly and set it aside for later. Use a soldering iron to connect the neutral (white) wire from the plug assembly to one banana socket. Connect the hot (black) wire from the plug assembly to the hot side of the outlet. Use the trimmed wire to connect the second banana socket to the neutral side of the outlet. Connect the ground wire (green) to the ground screw on the outlet. Use a multimeter to check the continuity of all connections. Secure the outlet inside the box and attach the cover plate.

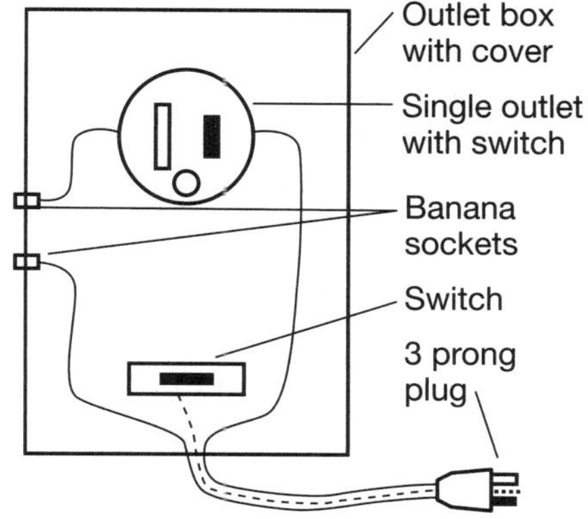

Ask students: Is there a limit to the number of appliances that one may connect to an outlet? Students will most likely provide anecdotes of times when they have "popped a fuse." Plug a power strip into an outlet and proceed to plug in appliances one at a time until the breaker on the power strip "pops." Ask students why the fuse popped. Discuss what increases as the appliances are added. In particular, ask students if the voltage and/or the current changes as the load on the circuit increases. Disconnect the appliances from the power strip and reset the breaker. Plug the power strip into the current interface and connect one multimeter (set to 20 amps AC) to the current interface and the other multimeter (set to 200 VAC) into the power strip, taking care not to touch the tips of the probes. Direct students' attention to the fact that the voltage reading equals about 120V while the ammeter will read zero. Add one appliance at a time to the circuit, each time stopping to measure the current and voltage output. As the appliances are added, the current increases, whereas the voltage remains relatively constant. Once the fuse is popped, direct students' attention to the fact that no current flows through the circuit. Discuss how power is the product of current and voltage in a circuit. Finally, talk about how the temperature of the wires in the circuit increases as the current increases, causing the breaker to pop.

Optional: Heat a bimetallic strip over the flame of a Bunsen burner to illustrate how the bimetallic strip inside a breaker bends as it is heated. As the strip bends it disconnects the circuit, emitting the familiar popping sound.

AC Blinkies

A neon bulb and an LED are used to differentiate between AC and DC current.

Application

AC current • AC versus DC current

Theory

Alternating current (AC) consists of current that is changing from positive to negative at a rapid rate. Electrical current in North America is delivered to consumers with the rate of 60 cycles per second: 60 positive pulses alternating with 60 negative pulses every second. A small neon bulb is used in this demonstration to show that household AC current is indeed pulsing at this rate. The neon bulb is a polarized device that glows only when current is flowing in one direction. Consequently, the bulb flashes on and off 60 times per second. When stationary, the bulb appears to have a constant glow since our eyes cannot detect flashes greater than a rate of approximately 20 cycles per second. If it is twirled rapidly in a circle, the on and off flashes of the neon bulb become evident. The bi-color LED, however, will glow continuously even when swung in a circle. This type of LED actually consists of two LEDs that are arranged so one LED glows when the current flows in one direction, while the second LED glows when the current flows in the other direction. When connected to an AC source, the LED will flash 120 times per second (60 red plus 60 green) and will appear as a mixture of the two colors when stationary. When twirled, the circle will appear as a chain of alternating colors. If connected to a DC source, this LED will glow only one color. The color depends upon the direction of the current.

Materials

Batteries, AA, 2

Battery holder, dual AA with alligator clip leads

Bulb, neon, small (RadioShack #272-1100B)

Hot glue or silicone sealant

Lamp cords with non-polarized plug, 4′–6′, 2

LED, bi-color (RadioShack #276-035), 2

Resistor, 40k Ohm, 1-watt

Resistor, 22k Ohm, 1-watt

Soldering iron and solder

Tape, electrical or heat-shrink tubing and heat gun

Wire cutters

Safety Precautions

Electrical Shock Hazard: Make sure that there are no exposed wires and that all insulation is in good condition. Follow all laboratory safety guidelines.

Wire the neon bulb in series with the 40-k resistor. Connect this combination across the ends of the lamp cord. Use hot glue or silicone sealant between wires to prevent short circuits. Solder all connections and wrap thoroughly with electrical tape or seal in heat shrink-tubing.

Wire one bi-color LED in series with the 22-k resistor. Connect this combination across the ends of a lamp cord. Use hot glue or silicone sealant between wires to prevent short circuits. Solder all connections and wrap thoroughly with electrical tape.

Demonstration

Plug the neon bulb assembly into a 110-V wall outlet. Ask students what they see. The lamp will glow, appearing constant. To show that the voltage is not constant, swing the bulb in a large circle in a darkened room. When twirled, the circle will appear as a chain of alternating on and off flashes. Discuss how the flashes demonstrate that the current is not continuous because the bulb is powered by AC current. Review how AC power consists of current that will change from positive to negative at a rate of 60 cycles per second.

Use a bi-colored LED and the battery holder to demonstrate that the LED glows green when current flows in one direction and red when it flows in the other direction.

Plug the other LED assembly into a wall outlet. It will probably appear to glow yellow or orange (a mixture of red and green). Tell students that this LED is exactly the same type that was previously connected to the battery assembly. Ask students why the LED does not appear red or green. Ask students what they expect to see if this LED assembly is twirled in a circle similar to what was done to the neon bulb assembly. Now swing this assembly in a circle in a darkened room and you will see alternating streaks of red and green light, showing that the current is in fact reversing direction 120 times each second (60 complete cycles).

Notes

Chapter 16

Magnetism

It All Started with Lodestones

The magnetic field of the Earth is mapped in a classroom using a series of magnets.

Application | Lodestones • Magnets • Magnetic field of Earth

Theory | The first reference to magnetism is found in a book called *The Book of the Devil Valley Master* (鬼谷子) that dates back to China in the 4th century B.C.E.: "the lodestone makes iron come or it attracts it." Lodestones are rocks made from magnetite, a naturally occurring ore of iron that formed in the presence of a magnetic field (the Earth's magnetic field). The name "magnet" is derived from lodestones found in Magnesia—an area located in Greece. According to the Dynamo Theory, the Earth's magnetic field is caused by the convection of molten (ferromagnetic) iron within the outer liquid core, along with the rotation of the Earth. This flow creates an electric current which is aligned along the north–south polar axis. The location of the poles is constantly changing due to the churning of the magma. In fact, geological records show that the poles have a tendency to flip every 50,000 to 1,000,000 years. The north pole of the magnetic needle of navigational compass points towards the (Magnetic) North Pole, indicating that the North Pole is in fact the south pole of the Earth's magnet.

Materials | Adhesive, construction

Bar magnet hangers, 12 (not essential)

Bar magnets, 13

Compass

Lodestone, large, preferably an elongated piece

Map, topographical, indicating geographical north and magnetic declination

Paper clips, 5–10

Paper correction fluid or white paint

Ring stand with ring

Tape, electrical

Thread, 12 m

Optional: Knitting needles (steel, 6″), 12, and a magnetizer

Safety Precautions | Although this demonstration does not pose any obvious risks, always follow laboratory safety rules. Keep magnets away from magnetic media including video and cassette tapes, computer floppy discs, credit and debit cards, and computer hard drives.

| **Preparation** | Use thread to suspend 12 bar magnets from the ceiling, no lower than 2 meters above the ground, all around the room. The magnets may be hung using magnet hangers, but this is not essential. Tying the thread to the middle of the magnet and securing with electrical tape, to prevent slippage, is satisfactory. The magnets should be spaced about 2 meters away from one another. Do not suspend the magnets near any large object made of iron or steel, since these objects may have become magnetized over time and therefore will interfere with the demonstration. Paint one end of the lodestone with paper correction fluid or white paint to serve as a reference point. An ideal piece of lodestone will have a point that will direct itself to magnetic North—paint this end white. Suspend the loadstone somewhere in the front of the room—perhaps from a ring stand on the demonstration table. |

Optional: Use a magnetizer to magnetize 12 six-inch steel knitting needles overnight. Suspend these as described above.

Demonstration

Direct students' attention to the lodestone hanging in front of them. Spin the lodestone gently and allow it to come to rest. Spin it again and when it comes to rest, ask students what they notice. Most should notice that the lodestone appears to come to rest in the same position—that is, pointing in the same direction. To dispel the suspicion that the thread forces the stone to hang a certain way, turn the ring stand base 90° and spin the stone again. Ask students if they know why the stone always appears to point in the same direction. Ask them if they can propose a use for such a stone.

Direct students' attention to the bar magnets hanging from the ceiling. Ask those nearest the magnets to reach up and gently spin the magnets. All the magnets will come to rest aligned with their north ends pointing towards magnetic north. Verify the direction using a compass.

Finally, show students a topographical map and explain how magnetic north moves over time. Show students how those who use topographical maps can compensate for this movement by calculating the magnetic declination. Discuss which pole of a magnet is at the "North Pole."

Reference

Li Shu-hua. "Origine de la Boussole 11. Aimant et Boussole;" *Isis,* vol 45, no. 2. (July, 1954), p 175.

Let's See the Lines

A magnet is inserted into a bottle filled with iron filings and oil, revealing a 3-D model of the magnetic field lines.

Application

Magnets • Magnetic field lines • Magnetic density

Theory

A field line of a magnet is a region in which a magnetic force is detected. The stronger the magnet, the greater the number of field lines present. Field lines spread out from one pole and move towards the opposite pole. By convention, the lines are said to move from the south to the north pole of the magnet. The magnetic field is greatest where the field lines are closer together. Regardless of the shape of the magnet, the field lines are closest at the poles. This demonstration uses two different types of cow magnets. The regular cow magnet has a simple field map that extends from one end of the magnet to the other. The multi-pole cow magnet, however, is made of 5–6 individual magnets connected end-to-end providing a more complex pattern. When they are grazing, cows have a tendency to pick up small pieces of metals as they eat. The metal can tear the digestive track of these animals and place them at risk of death. As a preventative measure, farmers have their cows swallow a strong magnet called a cow magnet that settles in the first stomach and serves to collect the shards of metal that the cow may ingest. During slaughter, magnets covered in shards of metal the size of a baseball have been found inside the stomach of cows.

Materials

Adhesive, construction

Cotton balls, 2–3

Cow magnets, one regular and one multi-pole

Dish, small

Iron filings, 10 cm^3 (about 70 g)

Mineral oil, 10–12 oz

Soda bottle, glass or plastic, colorless, 10–12 oz

Steel sphere, 1 cm

Test tube, 15 × 150 mm, with small lip

Preparation

Pour 10 cm^3 of iron filings into a *dry* soda bottle. Pour about 9 oz of mineral oil into the bottle and allow to sit for about an hour to permit bubbles to rise. Place the bottle on a small dish to collect any spills of mineral oil. Use a pencil to push 2–3 cotton balls into the test tube. Carefully slide the magnet into the test tube. The magnet will prevent the test tube from floating in the oil. Slowly insert the test tube into the bottle, allowing the excess oil to leak out. Lift the test tube and wipe its lip free of oil. Smear construction adhesive around the lip and push the test tube back into the bottle. Add more adhesive to create a permanent seal. Allow to set overnight. If the lip is small enough, the bottle cap may be reattached, if you choose.

Demonstration

Hold up the field demonstrator and gently swirl the empty bottle to disperse the filings. Once the filings have settled, slowly insert a regular cow magnet into the bottle. Once again, gently swirl the contents of the bottle. The iron filings will align themselves long the magnetic field lines of the magnet. Ask students to sketch the field lines in their notebooks and then to describe their arrangement. Ask students to identify the location of the poles and their relationship to the field lines. Discuss how the field lines are concentrated at the poles.

Show students the multi-pole magnet and ask them how the filings will distribute themselves around this magnet. Place the multi-pole cow magnet into the demonstrator and gently swirl the bottle. Ask students to sketch the field lines of this magnet in their notebooks and then to describe their arrangement. Ask students to identify the location of the poles.

Finally, pass a regular cow magnet and a small steel sphere around the classroom. Ask each student to compare how much force is needed to pull the sphere off of the sides of the magnet compared to pulling it off the poles of the magnet. Direct students' attention to the fact that the higher density of magnetic field lines is accompanied by a greater magnetic force.

Jumping Wire

A current-carrying wire is deflected by a permanent magnet.

Application | Magnetic field • Magnetic force • Current • Right Hand Rule

Theory | A wire carrying a current is accompanied by a magnetic field that encircles the wire. The direction of the magnetic field is summarized by the Right Hand Rule. To apply this rule: point your right thumb in the direction of the current; curl your fingers in a half-circle; your fingers will point in the direction of the magnetic field around the wire. As with all magnetic fields, the field lines travel from south to north. In this demonstration a wire is placed within the magnetic field of a permanent magnet. When a current is introduced into the wire, the magnetic field of the wire interacts with the magnetic field of the magnet, causing the wire to move. The direction of the motion is dependent upon the relative orientation of the poles of the permanent magnet. If the magnet is flipped, the direction of the field is reversed, resulting in the wire moving in the opposite direction. The wire may also be caused to move in the opposite direction by changing the direction of the current.

Materials | Magnet, horseshoe, 15 cm

Patch cord, 60 cm with alligator clips

Power source, DC, 10A, 12V

Safety Precautions | Although this demonstration does not pose any obvious risks, always follow laboratory safety rules. A wire connected directly to a power supply may overheat, potentially damaging the power source. In such cases introduce a current for only a few seconds.

Demonstration | Show students a patch cord and ask them what will occur within the wire if the cord is connected to a DC power supply. Discuss the current flowing through the wire, the temperature change due to the current, and the magnetic field due to the current. Ask students how they could demonstrate that a magnetic field exists in the current-carrying wire. After a short discussion place a horseshoe magnet on its side and drape the wire inside the magnet near the edge of the lower pole. Set the power supply between 9–12 volts with the switch in the off position. Connect the cord to the power supply and then turn on the power supply. After the wire jumps, shut off the power supply so that the cord does not overheat. Review the Right Hand Rule and discuss why the wire moved as it did. Ask students what can be done to the apparatus to cause the wire to move in the opposite direction. Test their hypotheses.

A model is used to simulate the operation of a mass spectrometer.

Application | Magnetic force • Mass spectrometer • Magnetic field

Theory | The masses of ionized particles can be measured with great precision in a device called a mass spectrometer. The ionized particles are accelerated into a magnetic field. The field forces each particle into a curved path. The radius of the path is dependent upon the mass and charge of the particle, as well as the strength of the magnetic field and rate of acceleration. The ions are directed towards various types of collection plates including CCD camera plates and photon counters. A mixture of different ions may be separated according to their mass to determine the relative composition of the mixture, creating a "fingerprint" of the mixture. The process may also be used to identify an unknown when compared to known samples passed through the mass spectrometer.

Materials | Adhesive, construction

Ball, non-ferrous, suggested ½″

Band saw, jigsaw or hand coping saw

Board, Masonite, ³⁄₁₆″, 24″ × 24″

Board, wooden 2″ × 4″, 10″ length (*Note:* A 2″ × 4″ actually measures 1½″ × 3½″)

Bottle cap, plastic soda

C-clamps, 5″, 2

Cups, plastic, 3-oz, 13 (Dixie® brand works well)

Drill

Drill bit, 1½″

Neodymium magnet, face-pole, ¾″ D × ½″ H

Ruler, 12″ plastic with a center groove

Steel balls of various sizes, suggested ⅛″–1″

Safety Precautions | Although this demonstration does not pose any obvious risks, always follow laboratory safety rules. Follow manufacturers' instructions when using power tools.

Preparation | Use a 1½″ drill bit to drill 10 holes along one side of the masonite board. Use construction adhesive to glue the open ends of 10 plastic cups over these holes. The cups will serve like pockets of a billiard table. Glue a cup in each of the other two corners and one in the center of the board. The 13 cups will serve to support the board like a table. Glue a magnet to the underside the board at a position of 12″ from the side opposing the row of cups and 6″ from one adjacent side. Glue a plastic soda bottle cap on top of the board to identify the location of the magnet. Use a band saw, jigsaw or hand-powered coping saw to cut an elliptically-shaped ramp from the $2'' \times 4''$. Use construction adhesive to attach a grooved plastic ruler to the slope of the ramp. Sandwich the ruler between the ramp and cut-off section of $2'' \times 4''$ and clamp together using two C-clamps to prevent the ruler from lifting as the adhesive sets.

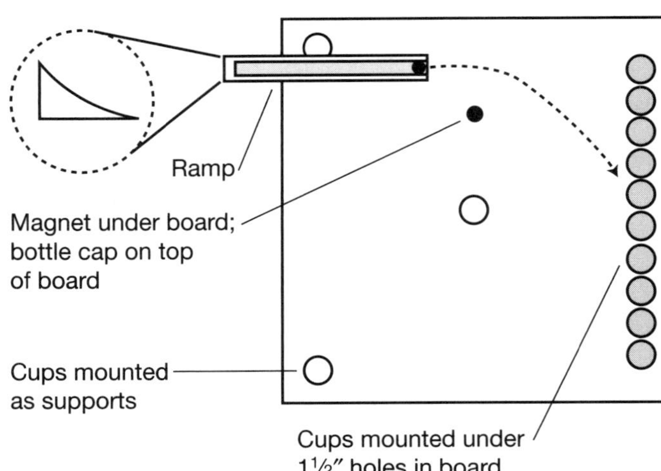

Demonstration

Review the operation of a mass spectrometer using this model. Use the ramp to launch a non-ferrous ball parallel to the sides of the board toward the line of cups. Direct students' attention to the cup in which the ball lands. This step simulates the spectrometer operating without a magnetic field. Without shifting the position of the ramp, launch the steel balls one at a time from largest to smallest. Note the relationship between the mass of a ball and the cup into which the ball lands.

Reference

Talesnick, Irwin. *The Idea Bank Collection—A Handbook for Science Teachers;* S17 Science Supplies & Services: Kingston, Ontario, Canada, 1984; Idea 51.

A current is passed through two parallel wires showing how the magnetic forces cause the wires to move.

Application | Magnetic forces • Right Hand Rule • Current

Theory | A wire carrying a current is accompanied by a magnetic field that encircles the wire. The direction of the magnetic field is summarized by the Right Hand Rule. To apply this rule: point your right thumb in the direction of the current and curl your fingers in a half-circle. Your fingers will point in the direction of the magnetic field around the wire. In this demonstration two current-carrying wires are positioned next to each other. The magnetic fields of the two wires interact with one another. When the current in the adjacent wires runs in the same direction, called parallel currents, the wires move together because the magnetic force is attractive. When the currents run in opposite directions, called anti-parallel currents, the wires move apart because the magnetic forces oppose one another.

Materials | Adhesive, construction

Board, masonite, $\frac{3}{16}''$, $9'' \times 24''$

Cup hooks, 2

Drill

Drill bit, $\frac{1}{2}''$

Dowel, wooden, $12'' \times \frac{1}{2}''$

Nails, common, $3''$, 2

Needle nose pliers with cutters

Patch cords with alligator clips, 3

Power supply, 10 amp, 12VDC

Right angle clamps, 2

Ring stands, 2

Wire, aluminum, $40'' \times \frac{3}{16}''$ (used for hanging ceiling tile grids)

Safety Precautions | Although this demonstration does not pose any obvious risks, always follow laboratory safety rules. Follow manufacturers' instructions when using power tools. A wire connected directly to a power supply may overheat, potentially damaging the power source. In such cases introduce a current for only a few seconds.

Preparation | Sketch a $3'' \times 18''$ box in the center of the Masonite board. This box will serve as a reference point. Drill $\frac{1}{2}''$ holes into each corner of the sketched rectangle. Attach a pair of cup hooks 3 cm apart over the top of the window. The screws of the hooks will protrude from the back of the board. Add a drop of construction adhesive to the inside of the collar of the hooks to permanently secure the hooks to the wood. Connect the protruding ends of the hooks using a small piece of aluminum wire. Use construction adhesive to glue two 3″ common nails horizontally about 2″ from the bottom of the window such that the heads are ¼″ apart. Allow the adhesive to set overnight. Mount the board between two ring stands using right angle clamps. Cut two pieces of aluminum wire 18″ in length. Use needle nose pliers to bend a hook at one end of each wire. Suspend these wires from the cup hooks such that the hanging ends are in contact with the nails. Adjust the angle of the board so the wires just barely touch the nails.

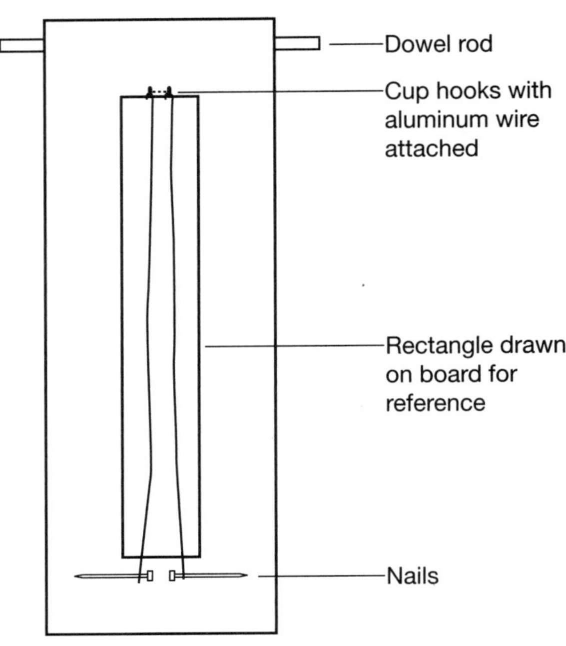

Demonstration

Present the demonstration device and show that the wires can freely swing from the cup hooks and slide across the nails. Ask students what will happen if a current is passed through both wires. Review the Right Hand Rule. Following discussion, connect the wires in parallel. First, connect the far ends of the nails with a patch cord. Then, connect one lead of the power supply to the connection between the cup hooks behind the board, and the other lead to one nail. Slowly increase the current in the wires and watch them move together. Once the wires move, shut off the power supply so the wires do not overheat. Ask students what can be done to the apparatus to cause the wires to move apart. To arrange the wires in an anti-parallel array, first remove the aluminum wire jumper connecting the nails. Then, connect one lead from the power supply to one nail and the other lead to the second nail. Slowly increase the current in the wires and watch them move apart.

A magnetic field is used to propel an aluminum pipe along a pair of conducting rails.

Application | Right Hand Rule • Magnetic field • Magnetic force

Theory | The Right Hand Rule is a simple way to determine the force acting on a current-carrying wire located in a magnetic field. In this demonstration an aluminum rod is placed on a pair of rails, which have a magnetic field directed perpendicular to the rails. A current flows from one rail to the other through the rod. The interaction between the current, the rod, and the constant magnetic field will cause the rod to be pushed down the rails. By knowing the direction of the current going through the rod and the direction of the magnetic field, one can predict the motion of the rod when a current is introduced.

Materials | Adhesive, construction

Angle irons, aluminum, $12'' \times \frac{1}{2}''$, 2

Board, wooden, $2'' \times 4'' \times 12''$

Board, wooden, $12'' \times 6'' \times \frac{3}{4}''$ (actual width = 5½")

Compass

Magnets, ceramic, flat, face-pole, enough to cover a surface area of $4'' \times 12''$ (Example: $1'' \times \frac{3}{4}'' \times \frac{3}{8}''$ ceramic magnets, 64)

Paint, white

Patch cords with alligator clips, 2

Pipe, aluminum, thin-walled, $\frac{3}{8}'' \times 6''$

Power supply, DC, 10 amp

Putty knife

Safety Precautions | Although this demonstration does not pose any obvious risks, always follow laboratory safety rules. Keep magnets away from magnetic media including video and cassette tapes, computer floppy discs, credit and debit cards, and computer hard drives.

Preparation | Use construction adhesive to glue the angle irons to the top face of the $12'' \times 6''$ wooden board, with the right angle facing towards the center of the board. Allow the adhesive to set overnight. Meanwhile, determine the south faces of the magnets using a compass (the south pole of a magnet will attract the north pole of the compass needle). Paint the south face of the magnets white. Use a putty knife to smear a thin layer of construction adhesive over the channel in between the two rails. Press the magnets into the adhesive with the south poles facing upwards. Place a $2'' \times 4''$ wooden board on top of the magnets to prevent the magnets from shifting and allow the adhesive to dry overnight.

Demonstration | Review the Right Hand Rule that describes the effect of the magnetic force on a current. Introduce the apparatus to students and draw attention to the direction of the magnetic field with respect to the rails. Connect the power supply to the rails and make students aware of the pole of each rail. Ask students what will happen if the aluminum pipe is placed on the rails in the center of the apparatus. Place the pipe on the rail and activate the power supply and watch the pipe move to one end of the system. Ask students what will happen if the poles are reversed and then repeat the demonstration with the poles reversed.

A coil of wire carrying a current is immersed in a magnetic field causing the coil to rotate.

Application | Magnetic force • Magnetic field • Right Hand Rule • Coils

Theory | When a current-carrying loop of wire is placed in a magnetic field, the loop will rotate. The rotation is due to the interaction between the surrounding magnetic field and the magnetic field accompanying the current in the loop. In this demonstration a coil of wire is suspended in a magnetic field provided by two strong magnets. The coil is initially suspended with its plane parallel to the field. When a current flows through the coil, the coil emits a magnetic field around the wire. The surrounding magnetic field is at a right angle relative to the coil and provides a net torque that causes the coil to rotate. Once the coil moves 90°, the torque acting on the coil is equal to zero and the coil no longer turns. The direction of rotation is dependent upon the direction of the current within the wire. This direction may be predicted using the Right Hand Rule. To apply this rule point your right thumb in the direction of the current and curl your fingers in a half-circle. Your fingers will point in the direction of the magnetic field in the wire.

Materials |
Adhesive, construction

Board, Masonite, $10'' \times 10'' \times {}^{3}\!/\!_{16}''$

Coffee can or similar cylinder, $4''$ diameter

Cup hooks, 2

Dowel, $14'' \times \frac{1}{2}''$

Drill

Drill bit, $\frac{1}{2}''$

Jigsaw or coping saw

Nail polish

Neodymium magnets, face-pole, $\frac{3}{4}'' \, D \times \frac{1}{2}'' \, H$, 2

Patch cords with alligator clips, 2

Power supply, 12VDC, 10 amp

Right angle clamps, 2

Ring stands, 2

Wire, aluminum, $3'' \times {}^{3}\!/\!_{16}''$ (used for hanging ceiling tile grids)

Safety Precautions | Although this demonstration does not pose any obvious risks, always follow laboratory safety rules. Follow manufacturers' instructions when using power tools. A wire connected directly to a power supply may overheat, potentially damaging the power source. In such cases introduce a current for only a few seconds.

Preparation

Draw a 6″ × 6″ square in the middle of the Masonite board. Drill a ½″ hole in each corner of this square. Use a jigsaw or coping saw to cut out the square. Use construction adhesive to glue a dowel to the backside of the Masonite frame as illustrated 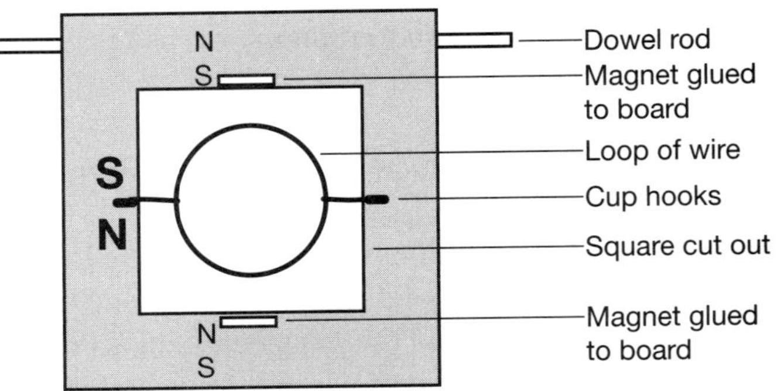 in the diagram. Insert a pair of cup hooks on either side of the square cutout from which the wire coil may be suspended. The screws of the hooks will protrude from the back of the board. Add a drop of construction adhesive to the inside of the collar of the hooks to permanently secure the hooks to the wood. Use construction adhesive to glue the sides of two magnets to the face of the Masonite frame. Arrange the magnets so their north poles face toward the top of the apparatus. Allow the adhesive to set overnight. Create a coil of 1½ loops of ³⁄16″ aluminum wire using a coffee can as a form. The ends of the wire coil should extend about 3″ from the coil, as illustrated in the diagram. Use nail polish as a paint to insulate any overlapping parts of the wire loop that touch one another. Since the loop overlaps on one side, a nonconductive counterweight must be added. Cut a piece of wire and form it into a semicircle identical to the radius of the loop. Coat this piece of wire entirely with nail polish. Use construction adhesive to attach this wire to the single-stranded side of the coil. Once the adhesive has set overnight, mount the apparatus from two ring stands. Suspend the coil from the cup hooks. Use a marker to indicate the poles of the magnets and the direction of the magnetic field across the coil.

Demonstration

Present the demonstration device and show that the coil can rotate freely on its axis. Direct students' attention to the labels, indicating the location of the poles of the magnets and the direction of the magnetic field across the coil. Ask students what will happen if a current is passed through the coil. Discuss how the direction of the coil's current affects its rotation. Use the Right Hand Rule to reinforce the relationship between current and the magnetic field of the wire. Connect the patch cords from a DC power supply to the protruding screws of the cup hooks. Quickly increase the voltage from 0 to 12 volts and observe the coil turn. Once the coil shifts, shut off the power supply so that the coil does not overheat. Ask students what can be done to the apparatus to cause the coil to move in the opposite direction. Test their hypotheses.

Field Lines of Coils and Solenoids

Iron filings are sprinkled on a coil of wire carrying a current revealing the field pattern of the coil.

Application Magnetic field lines • Magnetic forces • Solenoids • Electromagnets

Theory A field line of a magnet is a region in which a magnetic force is detected. The stronger the magnet, the greater the number of field lines present. Field lines spread out from one pole and move toward the opposite pole. By convention, the lines are said to move from the South to the North Pole. The magnetic field is greatest where the field lines are closest together. A current-carrying wire bent into a solenoid (coil) shape will display the same field map as a bar magnet due to the collective field of all the loops. If a piece of iron is inserted into a solenoid, the magnetic field increases greatly because the magnetic domains within the iron core amplify the magnetic field of the solenoid. A bundled core is more effective than a solid core of the same mass because the domains in the individual wires align more freely.

Electromagnets are superior to permanent magnets since the force field may be adjusted by changing the current or voltage. The strength of the electromagnet is determined by the size of the core of the solenoid. Electromagnets are found in electric motors, electromagnetic switches such as electric car door locks (solenoid switches), and in scientific research equipment whenever a precise or very large magnetic field is required.

Materials Bottle caps, plastic (from soda bottles), 4

Coat hangers, wire, 3

Drill

Drill bit, ³⁄₁₆″

Hot glue

Hot glue gun

Iron filings in a saltshaker

Magnet, bar, 15 cm

Marker, permanent

Patch cords with alligator clips, 2

Photo box frame, plastic, 8½″ × 10″, 2 (available from craft/dollar stores)

Pipe hangers, ½″, 2

Plexiglass® or Lexan®, 5″ × 7″ × ⅛″ piece

Tape, electrical

Wire, 12-gauge, insulated, solid core, 6′

Wire strippers/cutters

Power supply, 12VDC, 10-amp

Overhead projector

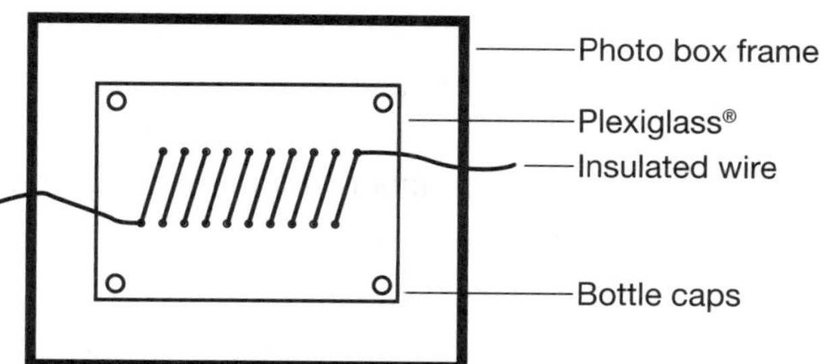

<table>
<tr><td>Safety
Precautions</td><td>Although this demonstration does not pose any obvious risks, always follow laboratory safety rules. Follow manufacturers' instructions when using power tools. A wire connected directly to a power supply may overheat, potentially damaging the power source. In such cases introduce a current for only a few seconds.</td></tr>
<tr><td>Preparation</td><td>Drill two lines of holes spaced about 1 cm apart in the plastic sheet, as illustrated in the diagram. One line of holes is shifted about 1 cm over in relation to the second line of holes. The lines should be centered along the length of the sheet about 5 cm apart. Thread the wires through these holes to form a coil that has a diameter of approximately 5 cm. Leave about 5 cm of wire protruding from either end of the coil. Use wire strippers to remove about 2 cm of the insulation from both ends. Use hot glue to mount a pipe hanger on the bottom of the coil board on either side of the coil along its center line. Cut 12 straight 20 cm pieces of wire from four wire coat hangers. Form a bundle of wires and wrap it with electrical tape to secure. The bundle should easily slide into the pipe hangers on the board. Glue four bottle caps to the bottom of the coil board. These caps will serve as feet to elevate the board. Place the coil board into an open plastic box frame. This frame will serve as a collection basin for excess iron filings that may damage the overhead projector. Use a permanent marker to sketch the outline of a bar magnet on the bottom of a second box frame.</td></tr>
<tr><td>Demonstration</td><td>Place the box frame with the outline of the bar magnet on top of the bar magnet on the stage of the overhead projector. Turn the overhead on to reveal a shadow of the magnet. Adjust the focus of the projector to project an image of the magnet. Ask students to sketch what they think the magnetic field lines of the bar magnet look like. Have students compare their hypotheses. Sprinkle iron filings into the plastic frame. Carefully lift the box frame upward, remove the magnet and return the frame to the overhead stage. Adjust the focus to reveal the field lines of the bar magnet. Have students compare what they observe to what they previously drew in their notebooks and make corrections if necessary.

Place the coil board and frame assembly on the stage of the overhead projector. Turn the overhead on to reveal a shadow of the coil. Adjust the focus of the projector to project an image of the coil board. Ask students to sketch what they think the magnetic field lines of a current-carrying coil of wire looks like. Have students compare their hypotheses. Sprinkle iron filings over the coil board. Connect the leads of the power supply to the ends of the coil and then increase the voltage to 8–12V. Tap the coil board to allow the filings to realign themselves within the magnetic field. Have students compare what they observe to what they previously drew in their notebooks and make corrections if necessary. Compare the field lines of the coil to the field lines of the bar magnet.

Carefully tilt the coil board and insert the bundle of wires into the pipe hangers. Once again, tap the coil board and observe how the field lines become more concentrated. Discuss the role of a core in a solenoid.</td></tr>
</table>

Chapter 17

Induction

What Good Is a Newborn Baby?

A magnet is brought near a coil of wire causing a current to flow.

Application   Induction • Electromagnetism • Current production • Coils

Theory   One day while working in his laboratory with two coils of wire connected to one another, Michael Faraday noticed that when a magnet was brought near one coil, the other coil deflected the needle of a compass situated nearby. The phenomenon fascinated Faraday and he soon showed his discovery to a number of people including British Prime Minister Robert Peel. In response to Faraday's enthusiastic demonstration, Peel remarked, "But, after all, what use is it?" Faraday then replied, "Why sir, there is the probability that you will soon be able to tax it." Peel then asked, "What good is it?" Faraday replied, "What good is a newborn baby?" Faraday foresaw that his discovery might spawn new frontiers.

Faraday's demonstration showed that when an electric conductor is in the presence of a moving magnetic field or if the conductor itself moves through a magnetic field, an electric current is induced in the conductor. This groundbreaking demonstration fostered the development of all electromagnetic devices including motors, generators, and transformers.

Materials   Adhesive, construction

Compass, plastic (must be see-through), e.g., Silva® brand

Dowel, wooden, $6'' \times \frac{1}{2}''$

Neodymium button magnet, 1-cm diameter

Overhead projector

Plexiglas®, $4'' \times 10'' \times \frac{1}{8}''$ (available at craft stores for photo frames)

Solid core wire on $\frac{3}{4}''$ plastic spool, 20-gauge, $25'$

Tape, electrical

Safety Precautions   Although this demonstration does not pose any obvious risks, always follow laboratory safety rules. Use caution when handling the neodymium magnet. Neodymium magnets are very strong and may quickly snap together or to other magnetic material and pinch skin. The magnet is fragile and may shatter if dropped or if it hits another object too hard. Keep the magnet away from computer disks, computer monitors, credit and debit cards, and televisions.

Preparation   Carefully unwind the wire from the spool without kinking it. Rewrap the wire onto the spool beginning $1'$ from the end of the wire so that, once wrapped, both ends of the wire extend from the spool. The other end of the wire should extend about $2'$ beyond the spool. The neater and tighter the windings, the more effective the coil will be. Wrap electrical tape around the spool of wire to prevent the coil from unwinding. Use construction adhesive to mount the coil on one end of the board so the core of the coil lies in a horizontal orientation. Allow the adhesive to set overnight. Wrap the longer lead of the coil around the compass four or five times and connect the two leads to one another. The wire leads may be soldered or simply twisted together. Place adhesive in the four corners of the back of the compass and then mount the compass to the board. Use adhesive to attach a neodymium magnet to the end of a dowel. Allow the adhesive to set overnight.

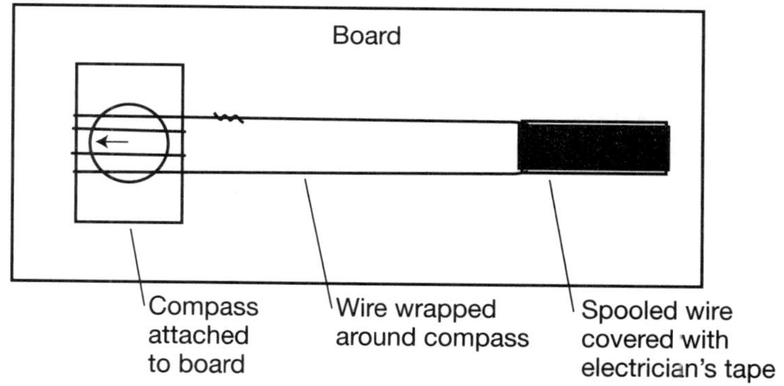

<table>
<tr><td>Demonstration</td><td>Place the demonstration board on an overhead projector and arrange it so the compass needle is in line with the wires wrapped around it. Insert the magnet in and out of the primary coil and observe how the compass needle deflects away from the wires. Ask students to explain why the needle is deflected when the magnet is inserted into the coil. Discuss how Michael Faraday used this groundbreaking demonstration as a stepping-stone to our understanding of electromagnetism.</td></tr>
<tr><td>Reference</td><td>Thomas, John Meurig. Michael Faraday and the Royal Institution: The Genius of Man and Place; CRC Press, Taylor & Francis Group: New York, 1991; p 128.</td></tr>
</table>

Tapping into the Earth's Magnetic Field

A long wire is whirled at a right angle to the Earth's magnetic field, resulting in a current flowing through the wire.

Application | Earth's magnetic field • Induction

Theory | Faraday demonstrated that when an electric conductor moves through a magnetic field, an electric current is induced in the conductor. In this demonstration, an extension cord that is connected to a galvanometer is twirled at a right angle to the Earth's magnetic field, inducing a current in the wire. The effect is most pronounced when the conductor is at a right angle to the magnetic field lines. If the conductor moves parallel to the field lines, a current is not produced. The magnitude of the field and the uniform distribution of the flux lines provide an opportunity to tap into the field to do useful work. Space scientists are presently perfecting techniques to power tethered satellites by tapping the Earth's magnetic field. The strength of the Earth's magnetic field varies from 30 to 60 microteslas (0.3–0.6 gauss), depending on location. According to the Dynamo Theory, the Earth's magnetic field is caused by the convection of molten (ferromagnetic) iron within the outer liquid core, along with the rotation of the Earth. This flow creates an electric current which is aligned along the north–south polar axis. The location of the poles is constantly changing due to the churning of the magma. In fact, geological records show that the poles have a tendency to flip every 50,000 to 1,000,000 years.

Materials | Extension cord, 50′

Extension cord plug, female

Extension cord plug, male

Galvanometer (preferably a demonstration galvanometer)

Insulated wire, 14–18 gauge, 1 m

Navigational compass

Patch cords, 6″, with one female banana plug lead and one free end, 2

Patch cords, 12″, with two male banana plug leads, 2

Screwdriver

Tape, duct

Tape, electrical

Wire strippers

Safety Precautions | Although this demonstration does not pose any obvious risks, always follow laboratory safety rules. Never connect the modified extension-cord plug to an outlet.

Preparation

Use a 30-cm piece of 14–18 gauge wire to connect the hot screw (thinner prong) of a male plug to the neutral screw (thicker prong) of the female plug. Use a 30-cm piece of 14–18 gauge wire to connect the ground screw (round prong) of a male plug to the hot screw the female plug. Connect a 30-cm patch cord to the neutral screws of the male plug. Connect a 30-cm patch cord to the neutral screw of the female plug. The plug adapters will provide a three-turn loop of wire when connected to an extension cord.

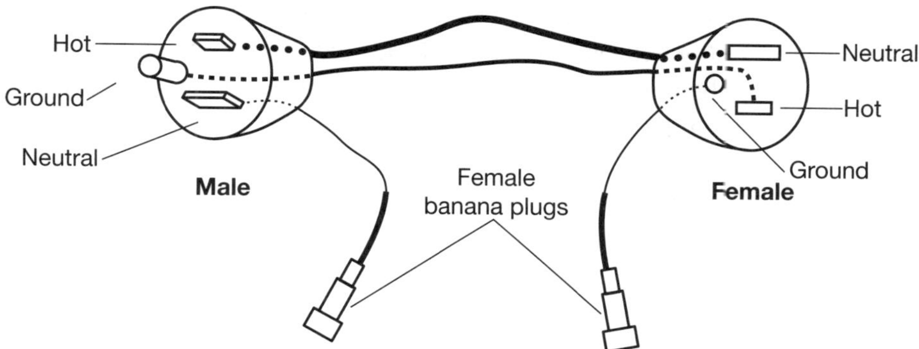

Connect the plug adapters to the extension cord. Attach the apparatus to a galvanometer using banana patch cords. Use duct tape to secure the cord to the demonstration table to prevent the galvanometer from falling in the event that the cords are tugged too hard.

Demonstration

Invite two students to twirl the extension cord like one would do when skipping rope. Be sure that the extension is oriented perpendicular to the Earth's magnetic field (an East–West alignment) without telling students why yet. As the students twirl the extension cord, direct their attention to the movement of the needle of the galvanometer. Ask students to explain why a current is produced when the wire is twirled. Have the volunteers change their orientation by 90° (now, North–South alignment) from their original positions without any mention of polar coordinates. Repeat the demonstration and point out that the current is reduced or hardly noticeable. Discuss what is needed to induce a current in a moving conductor. Finally, discuss how the Earth's magnetic field is strong enough to induce a current in any moving conductor.

Reference

Timmons, Daniel L. "Finding Magnetic North the Hard Way" *The Physics Teacher*; American Association of Physics Teachers: College Park, MD, 1988; vol 26, p 276.

Build a Simple DC Motor

A basic DC motor is constructed using a coil of wire, a battery and a magnet.

Application | Induction • Motors • Direct current

Theory | The first DC electric motor was invented by Anyos Jedlik in Hungary in 1828. It consisted of two electromagnets—one stationary and the other spinning. His motor was capable of propelling a toy vehicle. The motor built for this demonstration consists of an electromagnet that interacts with a permanent magnet. As current flows through the coil of the electromagnet, its magnetic flux opposes the field of the permanent magnet, causing the coil to rotate. As the coil rotates, its electrical connection is broken, causing the electromagnet to shut off. The coil continues to rotate due to its inertia, causing the coil to reconnect once again: on-off-on. The process continues making the coil spin smoothly. The speed of the motor may be increased by increasing the number of turns in the coil, the current through the coil, and the strength of the permanent magnet. Modern electric motors incorporate a commutator to reverse the current as the motor spins rather than using an on-off-on process. One type of commutator uses a split ring that changes the direction of the current as the coil connects from one side of the split to the other. A commutator doubles the efficiency of the motor since the force acting on the spinning coil is continuous.

Materials | Alligator patch cords, 2 or 3

Batteries, 1.5V, and holders, 2

Cup, plastic, disposable, any size

Cylinder, ½″ (e.g., pen, AA battery, or similar size object)

Emery board or fine sandpaper

Large paper clips, 2

Magnet wire, 24-gauge, 150 cm

Needle-nose pliers

Neodymium button magnets, ½″ D × ⅜″, 2

Tape

Safety Precautions | Although this demonstration does not pose any obvious risks, always follow laboratory safety rules. Use caution when handling the neodymium magnet. Neodymium magnets are very strong and may quickly snap together or to other magnetic material and pinch skin. The magnet is fragile and may shatter if dropped or if it hits another object too hard. Keep the magnet away from computer disks, computer monitors, credit and debit cards, and televisions.

A DEMO A DAY

Preparation

Use a ½″ cylinder to prepare two ½″-diameter coils of wire—one from 50 cm of wire and the other using a 100 cm length of wire. Secure each coil by wrapping the opposite ends of the wire around the opposite sides of the coil. Trim the protruding wire leaving about 2 cm on each end. The coil will rest on two saddles made from paper clips, as illustrated in the diagram. Use needle-nose pliers to bend each paper clip into the shape of the letter "Y." Use tape to secure the saddles onto the opposite sides of a plastic cup. Sandwich the bottom of the cup between two neodymium magnets. Suspend the arms of the coil in the saddles. The coil will arrange itself in a vertical position because one side of the coil has one extra ½-turn. With the coil in this position, identify the bottom half of the arms, which are resting on the saddles. Remove the coil and use an emery board to remove the coating on this side of the wire. Be sure to remove only this portion of the coating on both arms of the coil. Return the coil to the saddles. Test the motor by connecting the leads from a battery to the paper clip saddles. Prod the coil with a finger to start the motor spinning. Repeat the process to produce the second coil.

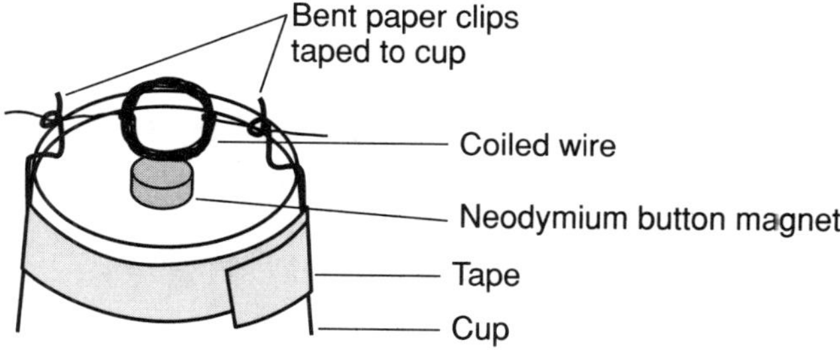

Demonstration

Prior to class, place the device on the demonstration table and activate the motor. (Students will no doubt notice the contraption as they enter the classroom.) Ask students if they can explain how the contraption works. Discuss the behavior and interaction of the spinning coil with the permanent magnet. Discuss how the magnetic field of the coil is intermittent due to the on-off-on electrical connection with the saddle. Compare this motor with an actual motor that uses a commutator to reverse the current flow, which will maximize the efficiency of the motor. Replace the small coil with the larger one and explore the effect the number of turns of wire has on the speed of the motor. Finally, explore the effect of voltage on the system by adding a second battery wired in series.

Eddy Current Race

A magnet slides down an incline made of aluminum at a slower than expected rate.

Application | Eddy currents • Lenz' Law • Induction

Theory | Faraday demonstrated that when an electric conductor is in the presence of a moving magnetic field, an electric current is induced in the conductor. The induced current also has an associated magnetic field surrounding the conductor. In 1834, Heinrich Lenz showed that these two magnetic fields have opposite signs. This is known as Lenz' Law. In this demonstration a magnet slides down an aluminum ramp. As the magnet moves across the aluminum, it induces an electric current within the surface of the aluminum. This current is known as an eddy current. As with any electric current, a corresponding magnetic field is present. According to Lenz' Law, the induced magnetic field and the magnetic field of the magnet oppose one another. Consequently, the movement of the magnet down the ramp is slower than expected. The eddy current effect is utilized in a number of ways, including in the separation of aluminum cans from a mixture of recyclables. Some roller coasters use the eddy current effect in their braking mechanisms.

Materials | Aluminum bar, $2'' \times 36'' \times \frac{1}{8}''$ or thicker (an aluminum T-square will work)

Coin, dime or quarter

Neodymium button magnet, $\frac{1}{2}''$ D

Safety Precautions | Although this demonstration does not pose any obvious risks, always follow laboratory safety rules. Use caution when handling the neodymium magnet. Neodymium magnets are very strong and may quickly snap together or to other magnetic material and pinch skin. The magnet is fragile and may shatter if dropped or if it hits another object too hard. Keep the magnet away from computer disks, computer monitors, credit and debit cards, and televisions.

Demonstration | Show students that the magnet is not attracted either to the aluminum bar or to the coin. Rest the bar on a stack of books or a box so the incline is about 30°. Hold the coin and the magnet in a flat position near the top of the incline and tell students you are about to release the objects down the ramp. Ask students to predict which object will reach the bottom of the incline first. After a short discussion, release the two objects simultaneously and observe how the coin shoots down the ramp, while the magnet slowly moves down the incline. Use Lenz' Law to explain the slowed motion of the magnet as it moves across the aluminum surface.

A series of coils is used to demonstrate how a transformer operates.

Application

Transformers • Coils • Magnetic flux

Theory

Transformers consist of two independent coils in which electrical energy is transferred from one to another using the process of induction. As current flows through one coil, the primary coil, it induces a current in the secondary coil. If the number of turns is equal in both coils, the amount of voltage and current is nearly the same, assuming that the coils surround a solid ferromagnetic core. The core allows for a greater transfer of energy between the coils. If the number of turns in the secondary coil, N_s, is greater than the primary coil, N_p, there is a proportional increase in the voltage (and a corresponding decrease in current):

$$\frac{N_s}{N_p} = \frac{V_s}{V_p} = \frac{I_p}{I_s}$$

Transformers are used to increase and decrease AC voltage. Household devices and appliances such as blenders, stereos, and vacuum cleaners use transformers to step-down the 120V of household current to various lower voltages. Television cathode ray tubes require a step-up transformer that provides about 15,000 volts. Transformers work only with alternating current since a constantly changing current is required to produce the magnetic flux. This demonstration uses the analog signal of a radio to provide a changing current.

Materials

Beaker, 250-mL or similar cylinder

Iron bar, solid, 2–3 cm (e.g., ring stand rod)

Phono plugs with solder connections, $1/8''$, 4

Soldering iron and solder

Speaker, audio amplifier (e.g., RadioShack Mini Amplifier Speaker, #277-1008)

Tape, electrical

Transistor radio, other electronic sound device

Wire cutters/strippers

Wire, 20–22 gauge magnet, 75′

Safety Precautions

Although this demonstration does not pose any obvious risks, always follow laboratory safety rules. Follow manufacturers' directions when working with soldering irons.

Preparation | Produce four coils of wire by wrapping the wire around a cylinder such as a beaker or mailing tube. Two coils should have 14 turns, one should have 7 turns, and the fourth should have, 21 turns. Use electrical tape to secure each coil so it does not unravel. The leads of each coil should protrude about 10 cm. Use a soldering iron to connect a $\frac{1}{8}''$ phono plug to each coil.

Demonstration | Connect a transistor radio or similar device to one 14-turn coil (the primary coil). Connect the other 14-turn coil (the secondary coil) to the input of the amplifier speaker. Turn on the sound source and slowly bring the coils together. As the coils come together, sound should be heard through the speaker. Discuss how the coils work together in producing the sound output. Ask students what will happen if the number of turns in the secondary coil is decreased or increased. Compare the volume of the sound using different combinations of the coils. Repeat the original demonstration using the two 14-turn coils, but this time insert an iron rod into the centers of the coils and show how the volume of the sound increases. Discuss the effect that the rod has on the flux of the coils.

References | Hobbs, Marsha. "Radio Transformer"; *PIRA News;* 1993; vol 6, No 3.

Rathjen, Don. "Modulated Coil"; *The Physics Teacher;* American Association of Physics Teachers, College Park, MD, 1998; vol 36, no. 7, p 416.

A frequency generator is used to power an AC motor.

Application | Alternating current • Induction • Motors

Theory | Serbian inventor Nikola Tesla built a motor in 1883 that utilized an alternating current to drive the spinning coil of the motor. This demonstration uses a square wave frequency generator to produce an alternating current in a coil, which is in proximity to a permanent magnetic field. The alternating current provides the alternating magnetic flux that causes the coil to rotate without interruption. Unlike a DC motor, a commutator is not required to reverse the flow of current in order to maximize the efficiency of the motor.

Materials | Coil apparatus from *Build a Simple DC Motor* demonstration, p. 308

Cylinder, ½″ (e.g., pen, AA battery, or similar size object)

Emery board or fine sandpaper

Needle-nose pliers

Patch cords with alligator clips, 2

Square wave frequency generator

Wire, 24-gauge magnet, 50 cm

Safety Precautions | Although this demonstration does not pose any obvious risks, always follow laboratory safety rules. Follow manufacturers' instructions when using power tools. A wire connected directly to a power supply may overheat, potentially damaging the power source. In such cases introduce a current for only a few seconds.

Preparation | Prepare a coil of wire from a 50-cm piece of wire as described in the *Build a Simple DC Motor,* p 308. Use an emery board to completely remove the coating on the arms of the coil.

Demonstration | Connect the patch cords from a DC power supply to the protruding screws of the cup hooks. Quickly increase the voltage from 0 to 12 V and observe the coil turn. Once the coil shifts, shut off the power supply so that the coil does not overheat. Ask students to explain why the coil does not continue to rotate as it does in the *Simple DC Motor* (see p 308). Direct students' attention to the fact that the coil in this apparatus is always connected to the cup hook saddles of the system. Discuss how the current in an electric motor is either intermittent or repeatedly reversed. Connect a square wave frequency generator to the saddles and slowly increase the wave frequency until the coil begins to spin.

Notes

Chapter 18

Nature of Light

There's More Than Meets the Eye

A phosphorescent sheet is used to reveal the invisible light on the edge of the visible spectrum.

Application | Electromagnetic spectrum • Ultraviolet light

Theory | The electromagnetic spectrum consists of a range of radiation, including radio waves, microwaves, infrared rays, visible light rays, ultraviolet rays, X-rays and gamma rays, ranging in wavelength from 10^3 to 10^{-12} m. Our eyes can see light in the range of only 400 to 700 nm. Our skin contains nerves that can sense the infrared radiation (heat) just beyond the lower range of the visible spectrum. In this demonstration a phosphorescent sheet is used as a detector of ultraviolet radiation that lies just beyond the violet range of the visible spectrum. Phosphorescent materials glow in the presence of light near the range of visible violet light and into the ultraviolet wavelengths. Although humans are limited to seeing the "visible spectrum," other creatures can detect other wavelengths. Snakes, including boas, pit vipers and pythons, are able to "see" using infrared-sensitive receptors found in the grooves between their nostrils and eyes or on their upper lip. This provides the snakes with the ability to see the radiated heat emitted from prey and other objects.

Materials | Diffraction grating (not 2D), 15 cm × 15 cm

File folder

Overhead projector

Phosphorescent sheet, 30 cm × 5 cm, or larger

Razor knife

Tape, masking

Preparation | Create a mask for the overhead stage by using a razor knife to cut a 10 cm × 5 cm window into the middle of a file folder.

Demonstration | Cover the stage of the overhead with the file-folder mask and project a horizontally-oriented window of light onto a wall. The projected window should be about 20 cm in width. Use masking tape to attach the diffraction grating to the outer lens of the projector. Explain to students that the grating acts as a prism by separating white light into its component colors. Two rainbows will be emitted—one on either side of the projected window. Use masking tape to mount the phosphorescent sheet so that one spectrum is centered over the sheet with a 5-cm space left beyond the edges of the visible spectrum. After about 30 seconds, cover the file-folder mask with a piece of paper. Direct students' attention to the glow that is visible on the phosphorescent screen. Move the paper on and off of the mask, flashing the projection of the spectrum on and off of the sheet. Ask students to look closely at the edge of the spectrum and explain what they see. Discuss how the phosphorescent sheet provides evidence that invisible light exists beyond the visible spectrum.

Use a solar cell and a video camera to detect the presence of invisible electromagnetic radiation.

Application	Electromagnetic radiation • Infrared radiation • Color filtration
Theory	The remote controls of most TV and electronic components use infrared wavelengths to transmit the signal to the device. The infrared region of the electromagnetic spectrum is not visible to our eyes. Infrared radiation, however, may be detected using a CCD video camera which makes the signal visible. This demonstration uses a solar cell to convert the infrared signal to an audible signal, allowing one to "hear the light." A violet filter will then be placed between the remote control and the camera to show how electromagnetic signals may also be blocked by absorption. In this case, the violet filter is made of a material whose atoms absorb the specific wavelength emitted by the remote control.
Materials	Gel filters, one each, red and violet
	Mini audio amplifier (e.g., RadioShack, # 277-1008)
	Solar cell with a mini phono plug
	TV remote control device
	Video camera and projector
Safety Precautions	Although this demonstration does not pose any obvious risks, always follow laboratory safety rules.
Demonstration	Point the remote toward students and press a few buttons and ask them if anything is emitted from the device. Discuss how a TV remote control works. Although the remote sends out a wireless signal, we cannot detect the signal with our senses. Point the controller at a video camera and activate the controller to show that it indeed sends a signal. Press different buttons to show how the flashing signal changes from button to button.
	Connect a solar cell to a mini sound amplifier. Discuss how the solar cell converts light energy into a current that then powers the speaker. Darken the room and shine a flashlight on and off a solar cell. The speaker will emit a sound pulse whenever the light flashes across the solar cell. Point the controller at the solar cell to "hear the infrared light signal." Press different buttons to show how the signal changes from button to button.
	Place a red filter between the remote and the camera to show that the red filter allows the transmission of infrared light. Replace the red filter with a deep violet filter to show that the violet filter blocks the infrared wavelength. This will confirm that the wavelength emitted by the remote is in the red–infrared range.

Are We Adding or Subtracting?

This demonstration differentiates between primary colors of pigments and primary colors of light.

Application

Primary colors of light • Primary colors of pigments • Filter

Theory

The colors blue (435.8 nm), green (546.1 nm) and red (700 nm) are known as the primary colors of light. Any color can be produced by mixing the correct proportions of these three colors of light. When all three colors of light are mixed in equal amounts, white light is produced. In this demonstration the primary colors of light are produced by passing white light through theater gel filters. A light filter is made of a material that allows the transmission of only one specific wavelength of light. The primary colors of light are often confused with the primary colors of pigment (reflected light). The primary colors of pigment are cyan (blue), magenta (red), and yellow. Pigments absorb specific wavelengths and reflect others. For example, a blue surface absorbs the red, orange, and yellow wavelengths but reflects the green, blue, and violet wavelengths, with the reflection appearing blue overall. In contrast, a yellow surface absorbs the blue and violet wavelengths of light but reflects red, orange, yellow, and green light. If the blue and yellow pigments are mixed, only the green light is reflected, with the other wavelengths being absorbed by the constituent blue and yellow pigments.

Note: The wavelengths of the primary colors of light were first defined by the International Commission of Illumination in 1931.

Materials

Compass

Flashlights, Maglite® with incandescent bulbs, 3

Gel filters: magenta, yellow, cyan

Gel filters: red ($\approx$ 680 nm), blue ($\approx$ 450 nm), and green ($\approx$ 130 nm)

Marker, permanent, fine-tipped

Overhead projector and screen

Pencil

Poster boards, 12″ × 12″, 3

Razor knife

Ring stand

Scissors

Tape

Utility clamps, 3

Safety Precautions

Although this demonstration does not pose any obvious risks, always follow laboratory safety rules.

 A DEMO A DAY

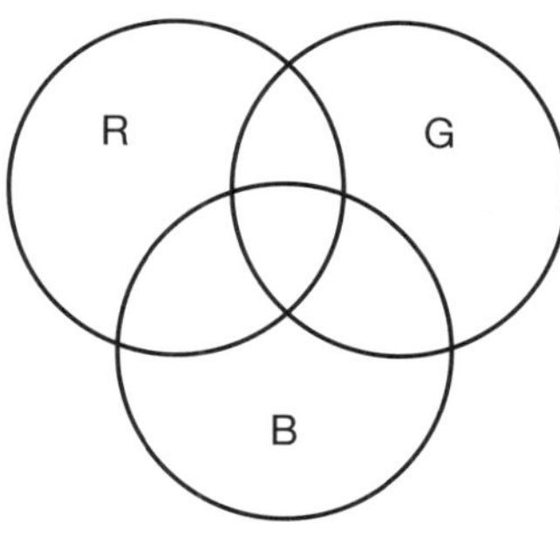

Primary colors of light when
mixed produce white light.

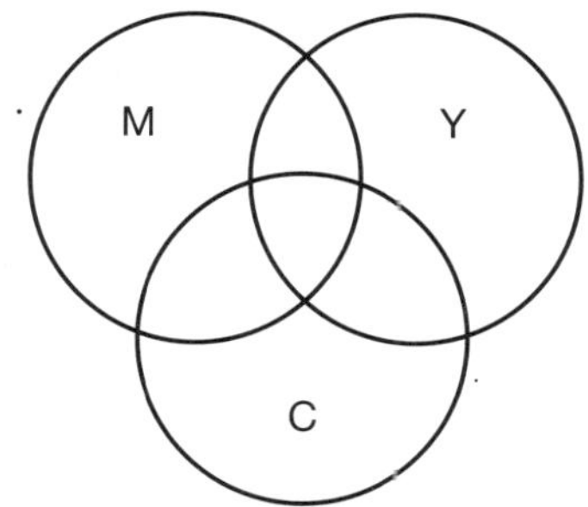

Primary colors of pigment when
mixed produce a black light.

Preparation

Remove a lens from a flashlight and use it to trace a circle from each of the red, blue, and green filters. Use scissors to cut out the circles. Sandwich one filter between the lens and ring of each flashlight and secure them to the bodies of the flashlights. Mount all the flashlights to one ring stand using three utility clamps. Use a compass and pencil to draw a 20-cm circle in the center of each piece of poster board. Use a razor knife to cut out these circles. Use tape to secure the filters over the three cut-outs.

Demonstration

Ask students to identify the three primary colors of light. Most likely students will not be in agreement since this concept is generally misunderstood. Project each colored flashlight onto a white surface, one at a time. Identify the colors of light as red, blue, and green—the primary colors of light. Explain that a filter is a material that allows the transmission of only one specific wavelength of light. Ask students what will happen if the colored beams of light are shone onto the same spot on the screen. Adjust the spots of light such that they overlap with one another, with all three overlapping in the center. The center will appear white since all three primary colors are reflected off a common surface. Remind students that a prism can be used to split white light into the colors of the rainbow. This demonstration does the opposite by combining the primary colors of light to produce white light.

Show students the magenta, yellow, and cyan filters and ask them what will be observed on the screen if the filters are layered on top of one another on an overhead projector. Remind students how a filter works. Place the filters on the stage of the projector so they overlap one another as illustrated in the diagram above. Ask students to explain why the center portion of the projection appears black.

Finally, review the difference between the primary colors of light and the primary colors of pigments. Discuss how artists mix pigments to produce specific colors. Compare this method to the method of producing colors of light on a video monitor.

Color Addition

Colored light of the three primary colors is added together to produce white light.

Application | Primary colors of light • Complementary colors • Reflection of light

Theory | White light can be dispersed into the elementary colors of the rainbow: ROY G BIV. These elementary colors of light may be recombined resulting in white light. Another way of producing white light is by combining the three additive primary colors of light: red, green and blue. These colors of light may be combined in various combinations and proportions to form all the colors of the visible spectrum. Red light combined with green light produces yellow light. White light is produced if the yellow light is also combined with blue light. The yellow light is considered to be a complementary color of blue, since it is the color that is added to the primary color to produce white light. Red light combined with blue light produces magenta. Blue light combines with green to form cyan.

Materials | **Method 1**

Beaker, 1-L, filled with ice water

Goggles

Light sticks, one each: red, blue, and green

String, 1 m

Tape, electrical

Method 2

Bolt, 2″

Drill, variable speed

Screwdriver

Washers, 2

Wing nut

Note: Depending on the type, the intensity of the light sticks may last all day. To prolong the life of the light sticks, however, store the them in ice water or place in a freezer between demonstrations. This will slow the rate of the chemical reaction.

Safety Precautions | Although this demonstration does not pose any obvious risks, always follow laboratory safety rules Wear impact-resistant goggles or safety glasses when performing this demonstration and be certain that the field of view of the audience is at a right angle to the motion of the apparatus.

Preparation | **Method 1:** Snap the light sticks to activate them. Lay them flat on a table side-by-side and tape them together at their ends, forming a rigid rectangle of the three cylinders. Tie a string securely to one of the end cylinders. Most light sticks have a hook and an eyelet. If the light stick does not have an eyelet, use an awl to make one. Tie a loop at the free end of the string large enough to slip your hand into.

Method 2: Snap the light sticks to activate them. Thread a washer followed by the three light sticks, followed by a washer and wing nut. Fan out the light sticks symmetrically. Tighten securely. Slide the apparatus into the chuck of the drill and tighten securely.

Demonstration

Ask students: What combination of light is necessary to form white light?

Method 1: Slip your hand into the loop of the string and swing the apparatus in a circle aligned in a vertical plane that lies perpendicular to the audience.

Method 2: Hold the drill in a position so the apparatus is in a circle aligned in a vertical plane that lies perpendicular to the audience.

To demonstrate the results of the addition of only two colors, use two light sticks at a time, arranging them opposite of one another. Repeat the procedure above.

Disposal

The used light sticks may be disposed of in the trash.

Can You See Me Now?

A polarizing filter is used to eliminate one of two images produced by a calcite crystal.

Application	Polarization • Electromagnetic waves • Index of refraction • Wave theory of light
Theory	According to the wave theory of light, light may be described as a transverse wave—similar to a vibrating rope. Like a vibrating rope, the light wave can travel in a vertical plane or horizontal plane. In fact, light waves travel in random planes. A polarizer may be used to organize light waves according to their orientation, blocking the transmission of 50 percent of the light. A picket fence may be used as a simplified analogy of a polarizing filter—only those waves of light that are more vertical than horizontal will pass through the narrow vertical slits of the fence. When two filters are combined such that their planes of polarization are at a right angle to one another, the combination results in all of the light being blocked. As the filters are rotated from 90° to 0° with respect to one another, the transmission of light increases from 0 to 50 percent.
	This demonstration utilizes a calcite crystal to produce a double image of a phrase. Calcite has the unique optical property of having two indices of refraction. A polarizing filter may be used to show how the light waves that form these two images are polarized 90° with respect to one another.
Materials	Calcite crystal, large, flat, smooth (with a face > 3 cm × 3 cm)
	Overhead projector
	Overhead transparency
	Polarizing filters, 5 cm × 5 cm, 2
Safety Precautions	Although this demonstration does not pose any obvious risks, always follow laboratory safety rules.
Preparation	Use a computer or permanent marker to print out a short phrase the size of the calcite crystal's face. For example: Can You See Me Now?
Demonstration	Place the transparency on the overhead and place one polarizing filter on top of the phrase. Project the image. Discuss how a polarizing filter eliminates one of the planes of the electromagnetic wave. Ask students: What will happen if a second polarizing filter is placed on top of the first filter? Discuss how the orientation of the second filter affects the transmission of the remaining plane of the wave. Rotate the second filter to show how all of the light may be eliminated when the two filters are at right angles to one another.

Now, remove the filters. Place the calcite crystal onto the phrase and rotate the crystal so the two images are most visible. Inform students that calcite has the unique property of having two simultaneous indices of refraction, thus producing two images. Hold one polarizing filter above the crystal and slowly rotate it. Students will notice that first image of the phrase will disappear and then reappear as the second image disappears. Discuss how this demonstrates that the two images produced by the calcite crystal are 90° out of phase with one another.

Why Is the Sky Blue?

Light from a projector is scattered to illustrate why the sky appears blue.

Application | Scattering of light • Transmission of light • Polarization

Theory

As sunlight passes through our atmosphere, the particles suspended in the air scatter the light in all directions. The process begins when the electrons in nitrogen and oxygen molecules absorb the energy (light) of the Sun. The electrons become excited and temporarily move to higher energy levels. This causes them to emit light in all directions, which is called scattering. The frequency of the scattered light is dependent upon the size of the particles in the air. When the air is "clean," it consists mainly of nitrogen and oxygen molecules. In this case, the light emitted by the electrons is in the blue part of the light spectrum. Thus the sky is blue. If the air contains smoke particles or water droplets, the electrons of these larger particles will emit light in the red region of the spectrum. The larger the particle, the lower the frequency of light that is scattered.

One may notice that the blue color of the sky is deepest at noon. In contrast, the sky appears red at sunset. At sunset the sunlight must travel a longer path through the atmosphere than at noontime. Consequently, the blue light is scattered to such an extreme that it becomes less visible than the transmitted red light. Thus the sky appears dimmer and red in color. In this demonstration a beam of light passes through a water solution of Pine-Sol®. The Pine-Sol imitates the scattering properties of our atmosphere by simulating both the transmission of red light and the scattering of blue light. A polarizing filter may be used to demonstrate that scattered light is polarized and transmitted light is not. Polarized sunglasses are designed to filter the light scattered by horizontal surfaces such as pavements, water surfaces, and the windshields of vehicles. Photographers utilize polarizing filters to diminish the glare of the atmosphere as well as reflective surfaces such as water and glass.

Materials

Aquarium, clean, 5–10 gal

Light, bright beam (a 35-mm slide projector is recommended)

Pine-Sol®, 300 mL

Polarizing filter, 15 cm × 15 cm, or larger

Poster board, white, 30 cm × 30 cm

Ring stand

Spoon, large

Water

Safety Precautions

Although this demonstration does not pose any obvious risks, always follow laboratory safety rules.

Preparation

Fill an aquarium ¾ full of water. If your source of water is hard, it is recommended that you use distilled or deionized water. Direct a bright beam of light through one side of the aquarium such that it is projected onto a piece of white poster board leaning against a ring stand. To provide the audience with the best view, the poster board should be oriented at a 45° angle with respect to the beam of light.

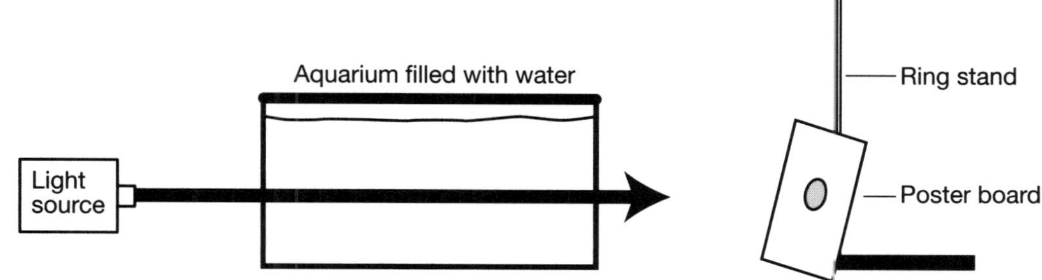

Demonstration

Turn on the beam of light and then darken the room. Direct students' attention to the apparatus and ask them to describe what they see. Students should easily see a spot of white light on the board. Some may notice a faint horizontal column of light in the water. Add about 100 mL of Pine-Sol to the water in the aquarium and use a spoon to gently stir the solution. Students should immediately notice that the column of light in the water becomes more visible. Furthermore, the column develops a blue tint. Students will also observe the spot on the screen developing a yellow–orange color which is dimmer than the original spot. Ask students to explain these changes. Discuss how the Pine-Sol filters the white-colored beam, revealing a yellow–orange spot on the board that is similar in appearance to a sunset. The Pine-Sol also scatters the blue wavelengths of light, causing the column within the water to appear blue in color. Compare this part of the demonstration to how the atmosphere scatters sunlight, making the sky appear blue in color.

Rotate a polarizing filter in between the projector and the aquarium to verify that the light source is not polarized. Next, determine if the transmitted beam is polarized or not by rotating the filter in between the aquarium and screen. Students will see that the filter has no effect on the transmitted beam. Finally, place the filter in between the aquarium and audience and rotate the filter to show that the scattered light is polarized. Discuss the advantage of using polarized sunglasses to diminish the intensity of scattered light.

Notes

Chapter 19

Optics

Is It Real?

A pane of glass is used to form a virtual image of a candle flame.

Application | Virtual image • Reflection • Photometry • Specular reflection • Law of reflection

Theory | Any smooth surface can exhibit specular reflection, which consists of a single incoming ray that is reflected in a single outgoing direction. Light from an object consists of many rays that reflect off a specular surface in an organized fashion, thus producing an image of the object. The reflected light seen by one's eyes is aligned with an image that appears behind the specular surface. This image is said to be virtual because the light from the object does not intersect behind the mirror. The image is not actually located behind the mirror, it only appears to be there. Plane mirrors can produce only a virtual image. A single plane mirror will produce a right–left inversion of the image.

Materials | Board, 24″ × 1″ × 3″, wooden

Candles, ¾″ × 4″, 2

Drill

Drill bit, ¾″

Matches or lighter

Tape, electrical

Table saw or hand saw

Window or photo frame glass pane, 5″ × 7″

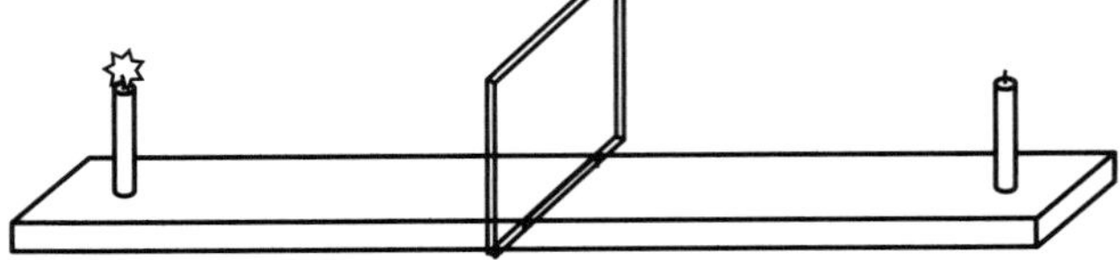

Safety Precautions | Although this demonstration does not pose any obvious risks, always follow laboratory safety rules. Follow manufacturers' directions when using power tools.

Preparation | Wrap electrical tape along the edges of the glass pane. Draw a line that bisects the exact center of the face of the wooden board. Use a table saw or hand saw to cut a ½″ groove along this line. The groove must be wide enough to snugly hold the glass pane. Drill two ¾″ holes located equidistant about 10″ away from either side of the glass pane. Mount the candles in these holes. For the best appearance, light the candles and allow them to burn for a few minutes or until they appear reasonably alike.

Demonstration | Position the candle device in a corner of the room or have students rearrange themselves so they are standing directly in front of the device. This will improve their vantage point. Align the device so the glass pane is facing the students. Light the candle located nearest the students and direct their attention to the other candle behind the pane. Students will notice that both candles appear lit. Place a finger into the "flame" of the unlit candle revealing that the flame is a virtual image. Discuss how the pane acts as a half mirror—reflecting the image of the candle flame according to the law of reflection.

A concave mirror is used to produce a real image.

Application | Real image • Mirrors • Apertures • Aberration

Theory

A concave mirror may be used to produce a real image of an object. A real image is one that can be focused onto a surface. Unlike a virtual image, the light rays of a real image intersect in an organized fashion on the surface. Unlike a virtual image, a real image can be projected onto a surface. If the object is placed between the focal point and the center of curvature of a concave mirror, the image will be real, inverted, and larger than the object.

The quality of the image cast by a curved mirror or lens depends upon the workmanship of the mirror or lens. An inferior mirror will produce a poorly focused image known as an aberration. The focus may be improved by limiting the amount of light received and emitted by the mirror by adjusting the aperture in front of the mirror. Decreasing the aperture eliminates some of the light that contributes to the image appearing out of focus. Although the image may now appear more in focus, it is also dimmer, showing that the improvement is not without consequence. The apertures of optical instruments are measured in terms of "f-stop numbers." The f-stop number is the ratio between the focal length of the mirror and the diameter of the aperture.

Materials

Adhesive, construction

Drawing compass

Gooseneck lamp with decorative torch lightbulb

Mirror, shaving or makeup, concave (15 cm diameter with a focal length of 50–100 cm)

Pencil

Permanent marker

Poster board

Razor knife

Right angle clamp

Ring stand

Tape, masking

Utility knife

Safety Precautions

Although this demonstration does not pose any obvious risks, always follow laboratory safety rules.

Preparation

Use construction adhesive to permanently mount a right angle clamp to the back of a large concave mirror. Allow the adhesive to set overnight. Cut four discs out of the poster board—each disc should have a diameter equal to that of the concave mirror. Using the following formula, calculate the aperture size for each of the following four f-stops: 1.0, 1.4, 2.0 and 4.0.

$$\text{Aperture size (cm)} = \frac{\text{focal length of concave mirror (cm)}}{\text{f-stop number}}$$

Use a compass to outline each aperture size in the center of one of the four discs. Use a utility knife to cut out the openings. Use a permanent marker to draw a large "F" on the lightbulb.

Demonstration

Mount the mirror on a ring stand. Dim the lights and draw the shades in the room. Place the lamp in front of the mirror at a position between its focal length and center of curvature and project an image of the lightbulb with the letter "F" onto a wall. Discuss the properties of a real image and what is required to produce one. Direct students' attention to the quality of the image in terms of sharpness, contrast, and focus. Ask students what can be done to improve the quality of the image. Cover the mirror with the f-stop = 1.0 aperture to show that the image may be improved by limiting some of the light reflected off the mirror. Use masking tape to secure the apertures on the mirror. Add successive apertures to show that the image may be further improved, but not without consequence. Although the sharpness of the image improves, the image becomes dimmer as the aperture size decreases. Discuss how apertures are used in optical instruments such as cameras, telescopes, and microscopes.

Reference

Carpenter Jr., D. Rae; Minnix, Richard B. *"The Dick and Rae Physics Notebook"*; Dick and Rae: Lexington, VA, 1993; O-170.

Flying Physics Teacher

A mirror is used to make a person appear to levitate.

Application	Reflection • Virtual image
Theory	When an object is placed in front of a plane mirror, an image of the object appears behind the mirror. This image is said to be virtual because the light from the object does not intersect behind the mirror. The image is not actually located behind the mirror—it only appears to be there. Unlike a real image, a virtual image cannot be projected onto a surface. Plane mirrors can only produce virtual images. A single plane mirror will produce a right–left inversion of the image. When straddling a mirror, a virtual image of the leg in front of the mirror appears behind the mirror with right–left inversion, which is why the image looks like the other leg.
Materials	Mirror, square, 24″–30″ Ring stands, large, 2 Step stool or small ladder Utility clamps, 2
Safety Precautions	Although this demonstration does not pose any obvious risks, always follow laboratory safety rules. Use caution when climbing on and off a lab bench.
Preparation	Position the mirror vertically in the middle of a lab bench using two ring stands and utility clamps. Face the mirror toward the audience.
Demonstration	Use a step stool or small ladder to climb onto the lab bench. Straddle the mirror so it is between your legs. Bend your knees slightly to improve the illusion and begin flapping your arms like a bird. Slowly lift the leg in front of the mirror while continuing to flap. You will appear to levitate. Students will most likely recognize how the illusion is achieved. Ask students exactly where the mirror must be located to be most effective. Discuss the relationship between the object distance and the distance of the virtual image.
Reference	Zwicker, Earl. "Flying Halloween Witch"; *The Physics Teacher;* 1988; vol 26, p 476.

Follow the Ray

A fog machine is used to illustrate the ray diagram of a curved mirror.

Application

Ray diagrams • Mirrors • Lenses • Tyndall effect

Theory

Lasers are used in this demonstration to allow us to envision the path light takes to and from a concave shaped mirror. A beam traveling along the principal axis is reflected back along the same path. A beam traveling parallel to the axis is reflected through the focal point of the mirror. A beam that travels through the focal point is reflected parallel to the principal axis. Fog is used to make the laser beams visible as they travel to and from the mirror since the fog particles are large enough to reflect light. This process relies on the phenomenon known as the Tyndall effect, which is the basis of Rayleigh scattering. This demonstration uses red and green lasers to differentiate the three beams examined. Students will notice that the green laser appears brighter than the red laser despite the fact that they both have a power rating of 5 mW. The apparent difference in brightness may be explained by the fact that human eyes are more sensitive to green wavelengths, making the green beam appear brighter. The technique of using a fog machine (or similar source of fog) to illuminate laser beams may be employed whenever laser beams are used to demonstrate any type of ray diagrams.

Materials

Clamps, adjustable angle or adjustable utility, 3

Fog machine (ultrasonic humidifier or Fantasy FX™ Fog in a Can)

Laser pointers, green, 2

Laser pointer, red

Mirror, large, concave, $f \geq 15$ cm, (e.g., a mirage mirror, available from novelty stores)

Right angle clamp

Ring stands, 2

Safety Precautions

Although this demonstration does not pose any obvious risks, always follow laboratory safety rules. Never look directly into the beam of a laser or direct the beam towards anyone's eyes. Do not use lasers that exceed 5 mW in power.

Preparation

Use construction adhesive to permanently mount a right angle clamp to the back of a large concave mirror. Allow the adhesive to set overnight. Mount the mirror on a ring stand. Use adjustable angle clamps or utility clamps to mount the three lasers to a second ring stand, with the red laser placed nearest the bottom. If necessary, use a drill to bore out the hole in the adjustable angle clamps. The screws of the clamps may be used to turn the lasers on and off. Direct the red laser towards the concave mirror along its principal axis. The second laser must be adjusted so it enters the mirror parallel to the axis and is then reflected through the focal point. Direct the third laser through the focal point before it is reflected off the mirror. Prior to demonstration verify that the beams are traveling through the focal point of the mirror and adjust, if necessary.

Demonstration Place the apparatus in front of students with the laser beams in the off position and directed toward the mirror facing the side of the room. (Lasers should never be directed toward the audience.) Ask students to predict the path of each laser beam. Following discussion, use a fogger to fill the area surrounding the apparatus with some fog. Turn on one laser at a time to verify students' predictions. To conclude the demonstration, turn on all three lasers at the same time.

Reference Thanks to William Koenig of Pascack Valley High School, Hillsdale, NJ, for sharing this idea.

Inside Up, Outside Down

A pendulum bob oscillates in and out of the focal point in front of a concave mirror.

Application	Concave mirrors • Focal point • Virtual image
Theory	In this demonstration, a pendulum oscillates toward and away from a concave mirror. The pendulum bob is painted so the top hemisphere may be differentiated from the bottom. As the bob swings in and out of the focal point, F, the virtual image in mirror changes from inverted (beyond F) to right side up (inside F). No image is formed when the bob is located at the focal point. As the bob oscillates, the image also changes in size. When the bob is located beyond the focal point, the image is larger than the object. When the bob is inside the focal point, however, its image is smaller than the object.
Materials	Adhesive, construction Ball, 1″ rubber (e.g., Superball®) Dowel, wooden, ½″ × 10″ Fishing line, colorless Mirror, large, concave, focal length (f) ≥ 15 cm, (e.g., a mirage mirror, available from novelty stores)* Paint (a color that contrasts with the color of the ball) Paint brush, small Right angle clamps, 2 Ring stand *Note: The focal length, f, measures the distance from the center of the lens to the focal point, F.
Safety Precautions	Although this demonstration does not pose any obvious risks, always follow laboratory safety rules.
Preparation	Use construction adhesive to permanently mount a right angle clamp to the back of a large concave mirror. Allow the adhesive to set overnight. Mount the mirror on a ring stand. Paint half of the rubber ball with a contrasting colored paint. Allow the paint to dry overnight. Use a right angle clamp to secure a wooden dowel above the mirror. Measure the distance between the dowel and the center of the mirror. Make a pendulum bob that corresponds to this distance from the rubber ball, fishing line and construction adhesive. Mount the ball so the equator between the two colors lies in a horizontal plane. Adjust the suspended bob such that it rests exactly at the focal point of the mirror.

 —

Invite students to gather in front of the mirror–pendulum system. Direct students' attention to the bi-colored pendulum bob. Ask students to describe what they see in the mirror. Since the bob is located at the focal point of the mirror, no image will be present. Oscillate the pendulum toward and away from the mirror and ask students to describe and explain what they see. Students should see the image of the bob appear to flip as it moves through the focal point.

Alternatively, the device may be left on a table for students to explore on their own followed by a discussion in class the following day.

Where Did It Go?

A beaker is made to "disappear" by submerging it in oil.

Application

Index of refraction • Refraction • Reflection

Theory

When we look at an opaque object, the object is visible because light reflects off its surface. A transparent object is visible because the light that travels through it bends (some reflection also contributes to its visibility). In this demonstration one glass beaker is placed into another. If we imagine light rays originating from behind the beaker system, the rays enter the outer beaker, bend toward the normal as they travel from air into glass, and then bend away from the normal as they travel from glass into air. Although the light beam exits the glass at the same angle that it entered the glass, the beam shifts. The shift of the light rays in combination with reflected light makes the transparent object visible. When the small beaker is immersed in oil, the air–glass interface is replaced by an oil–glass interface. The vegetable oil has an index of refraction that is nearly identical to the index of refraction of borosilicate glass. Consequently, there is little or no refraction of light as the beam travels through the small beaker. Since there is no difference between the indices of refraction of the glass and oil, no reflection occurs at the oil–glass interface.

Materials

Beaker, borosilicate glass, clean, 250-mL

Beaker, borosilicate glass, clean, 600-mL

Vegetable oil (any brand), 1 qt

Safety Precautions

Although this demonstration does not pose any obvious risks, always follow laboratory safety rules. Any food-grade items brought into the lab are considered laboratory chemicals and should not be consumed or removed from the lab.

Demonstration

Place the smaller beaker inside the larger one. Ask students if they can see both beakers. Discuss how light travels through the beakers as well as being reflected off them. Pour vegetable oil into the smaller beaker and discuss what the students see. Continue pouring oil until the oil overflows the smaller beaker and completely submerges it inside the larger beaker. As the oil is being poured, ask students to describe what they see. Once the system is filled, discuss why the beaker is no longer visible.

Disposal

Pour the oil back into its original container and save for next time. Wash the glassware with soap and water.

The disappearance of a piece of glass is investigated using a laser beam.

Application | Index of refraction • Refraction • Reflection

Theory

This demonstration is a follow-up to the demonstration "Where Did It Go?" on the previous page. In this demonstration, a laser beam travels through a glass plate that is situated inside a glass container with parallel sides. The glass plate causes the beam to shift because the beam is refracted twice—once when it enters the air–glass interface and again when it leaves the glass–air interface. The laser beam also reflects off the surface of the glass plate at the air–glass interface. When oil is added to the jar, the air–glass interface is replaced by an oil–glass interface. The vegetable oil has an index of refraction that is nearly identical to the index of refraction of borosilicate glass. Consequently, there is little or no refraction of light as the beam travels through the glass plate, causing the laser beam to shift away from its original position. Furthermore, since there is no difference between the indices of refraction of the glass and oil, no reflection occurs at the oil–glass interface.

Materials

Battery jar, square, clean glass, 1-L

Laser pointer, green

Modeling clay, small lump

Plate glass, small piece (e.g., ¼″ × 3″ × 3″)

Right angle clamp

Ring stand

Tape, masking

Vegetable oil (any brand), 1 qt

Safety Precautions

Although this demonstration does not pose any obvious risks, always follow laboratory safety rules. Never look directly into the beam of a laser or direct the beam toward anyone's eyes. Do not use lasers that exceed 5 mW in power. Any food-grade items brought into the lab are considered laboratory chemicals and should not be consumed or removed from the lab.

Preparation

Use a small piece of modeling clay to support a piece of upright glass protruding at a 45° angle from one corner of a square battery jar. Use a right angle clamp to mount a laser pointer to a ring stand. Direct the laser beam through the glass plate so the beam enters the jar at a right angle. Arrange the apparatus so the beam lands on a wall.

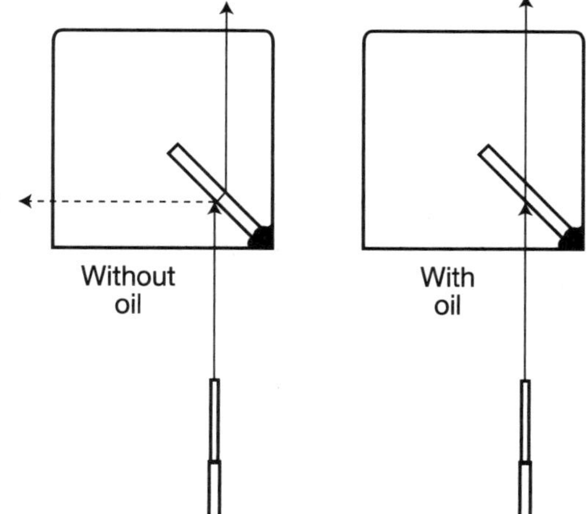

Demonstration | Ask students if they can see the glass plate inside the jar. Discuss how light travels through glass as well as being reflected off it. Turn on the laser and dim the lights. Use masking tape to mark the locations of the spot produced by the refracted beam as well as the spot produced by the reflected beam. Ask students what will happen if vegetable oil is poured into the jar. Slowly pour vegetable oil into jar. Once the laser beam is submerged, the refracted spot is no longer refracted and the reflected beam no longer exists.

Disposal | Pour the oil back into its original container and save for next time. Wash the glassware with soap and water.

Lasers are used to illustrate the refraction and reflection of light rays.

Application | Reflection • Refraction • Snell's Law

Theory | In this demonstration two laser beams are used to illustrate the path light takes as it travels between two media. A red laser travels from air into water at a right angle toward the surface of the water (Point A). Since the angle of incidence is zero, this beam is not refracted but simply transmitted into the water. The air is laced with fog and the water contains a minute amount of dispersed fat molecules to improve the visibility of the laser beam. The red laser also serves as a reference (called the normal line) for the green laser. Since the green laser enters the water at an angle, the beam is refracted. The beam refracts towards the normal since the beam is moving from a low index of refraction ($n_{air} = 1.00$) to a relatively high index of refraction ($n_{water} = 1.33$). At Point A, a percentage of the green laser is also reflected—following Snell's Law. Furthermore, the green laser reflects off the mirror lying at the bottom of the aquarium (Point B)—also following Snell's Law. The beam then refracts away from the normal as it goes from a high index of refraction to a relatively low one. Point C, the exit point of the green laser, is equal to the incident angle at Point A. Consequently, the reflected beam from Point A is parallel to the refracted beam at Point C. There is some reflection of the beam at Point C off the inner surface of the water. This beam will be parallel to the incident beam at Point A. A protractor may be used to compare the incident rays to the reflected and refracted beams.

Materials | Aquarium, glass, 10-gal

Clamps, adjustable angle or adjustable utility, 2

Coffee creamer, powdered, pinch

Fog machine (ultrasonic humidifier or Fantasy FX™ Fog in a Can)

Laser pointer, green

Laser pointer, red

Protractor, acrylic, full circle, 2

Ring stands with cross bar, 2

Right angle clamps, 2

Water, 7 gal

Safety Precautions | Although this demonstration does not pose any obvious risks, always follow laboratory safety rules. Never look directly into the beam of a laser or direct the beam toward anyone's eyes. Do not use lasers that exceed 5 mW in power.

Preparation

Pour 7 gallons of water into a 10-gallon glass aquarium. Dissolve a pinch of powdered coffee creamer in the water. Use right angle clamps to mount crossbars on each ring stand in a horizontal fashion above the aquarium. Use adjustable angle clamps or utility clamps to mount a laser to each crossbar. If necessary, use a drill to bore out the hole in the adjustable angle clamps. The red laser must be oriented downward, perpendicular to the surface of the water over the aquarium—about $\frac{1}{3}$ the distance from its edge. Mount the green laser and direct its beam toward the point where the red beam enters the water. Adjust the beams so the green beam reflects off the mirror placed in the bottom of the aquarium.

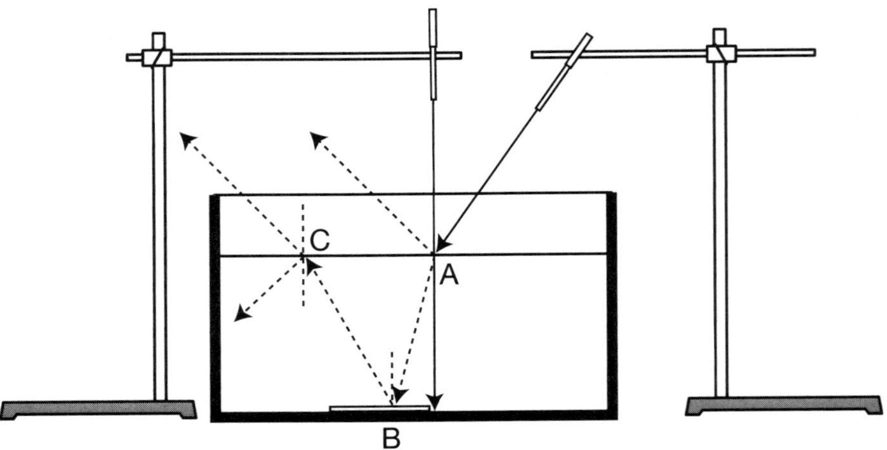

Demonstration

Invite students to gather in front of the aquarium. Inject a puff of fog over the water in the aquarium. Turn on the red laser and ask students to describe the path of the beam before and after it enters the water. Turn on the green laser and ask students to describe its path. Discuss the refraction and reflection of the green beam as it travels into, through, and out of the water. Add fog as needed. Use protractors to compare the incident and refracted beams at Points A and C.

A laser is used to reveal the critical angle needed for total internal reflection.

Application	Critical angle • Total internal reflection • Refraction
Theory	When light travels from one medium toward a second medium with a lower index of refraction, the light beam may exhibit total internal reflection. Whether or not the beam reflects depends upon the critical angle of the combination of media, since $\sin\theta_c = n_2/n_1$. Total internal reflection can occur only if $n_1 > n_2$. Diamond cutters cut diamonds in a way to promote total internal reflection, resulting in nearly all of the light exiting through the most visible facets of the diamond. Fiber-optic cables also utilize this principle to minimize the loss of light as it travels through the cable, allowing the signal to travel greater distances.

Materials

Adjustable angle clamp

Aquarium, glass, 10-gal

Coffee creamer, powdered, pinch

Fog machine (ultrasonic humidifier or Fantasy FX™ Fog in a Can)

Laser pointer, green

Ring stand

Water, 7 gal

Safety Precautions

Although this demonstration does not pose any obvious risks, always follow laboratory safety rules. Never look directly into the beam of a laser or direct the beam toward anyone's eyes. Do not use lasers that exceed 5 mW in power.

Preparation

Pour 7 gallons of water into a 10-gallon glass aquarium. Dissolve a pinch of powdered coffee creamer in the water. Mount a laser to a ring stand using an adjustable angle clamp. Position the beam at the bottom of one side of the aquarium so the beam is directed

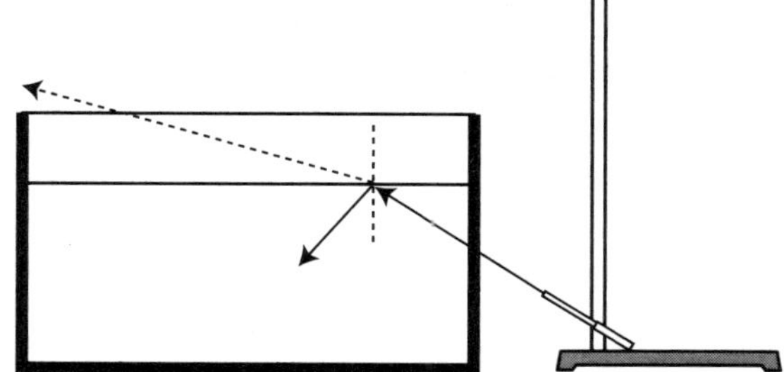

through the top surface of the water. Adjust the beam so that it is initially lower than the critical angle. Arrange the apparatus so students will obtain a side view with the spot of the reflected beam visible on the ceiling.

Demonstration

Inject a puff of fog over the water in the aquarium. Turn on the laser and ask students to describe the path of the beam. Discuss the angles of the reflected and refracted beams with respect to the water–air interface. Slowly increase the angle of the laser as measured with respect to the normal until total internal reflection is achieved. Direct students' attention to the fact that the spot on the ceiling disappears along with the refracted beam. The disappearance of the refracted beam is accompanied by the increase in brightness of the reflected beam.

Light Pipe

The principle of optical fibers is demonstrated using a bottle of soda.

Application | Reflection • Total internal reflection • Fiber optics

Theory | An optical fiber, often called a light pipe, is used to transport light. An optical fiber is made from long, thin flexible fibers of high purity glass or plastic, which is coated with a type of glass or plastic that has a lower index of refraction than the fiber's core. Optical fibers use the principle of total internal reflection to transfer the light beam from one end of the cable to the next. Light entering the fiber will reflect back and forth along the length of the fiber until it exits at its end. The difference between the indices of refraction of the core and the coating of the fiber makes the total internal reflection of the light possible. Optical fibers are used by worldwide telecommunication companies. They are also used in medicine to channel surgical laser beams and for viewing the inside of the body. This demonstration shows that a stream of water flowing from a bottle can be used as an optical "channel" to transport a laser beam.

Materials | Awl or nail

Laser

Shallow pan (cookie sheet)

Soda bottle, 2-L

Tape

Water

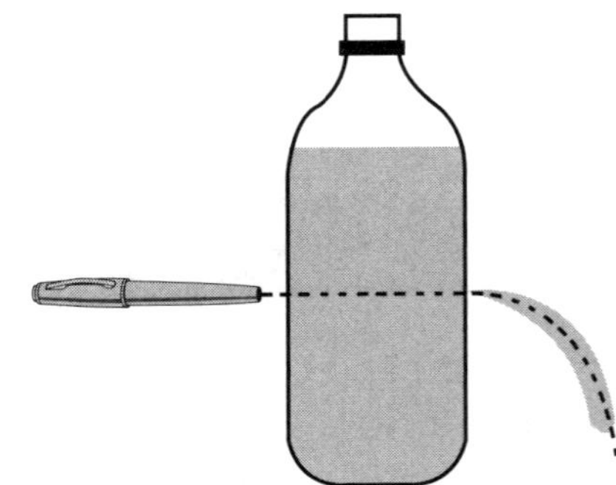

Safety Precautions | Although this demonstration does not pose any obvious risks, always follow laboratory safety rules. Laser light can cause permanent damage to eyes. Do not direct a laser beam toward anyone's eyes. Do not use lasers that exceed 5 mW in power. Clean up spills immediately.

Preparation | Use an awl or small nail to punch a hole into the side of a 2-L soda bottle, about 10 cm from the bottom. Cover this hole with a piece of tape. Fill the bottle with water.

Demonstration | Place the soda bottle on a cookie sheet. Turn down the lights and direct a laser beam through the soda bottle aimed at the hole. Remove the tape from the bottle to allow the water to flow. The laser beam will travel through the stream pouring from the bottle.

Disposal | Recycle the soda bottle.

Reference | Bilash, Borislaw. *A Demo a Day—A Year of Physical Science Demonstrations;* Flinn Scientific: Batavia, IL, 1997; p 314.

Playing cards are placed on the surface of an aquarium to illustrate total internal reflection.

Application | Total internal reflection • Index of refraction

Theory | Playing cards are mounted on the outer surface of an aquarium filled half-way with water. Card A is mounted on the bottom of the tank. It is visible only from the top of the water surface. Card A is not visible from the sides of the aquarium because the angle of the light reflected off this card exceeds the critical angle with respect to the sides of the aquarium. The light reflected off the card therefore exhibits total internal reflection. The refraction of light reflected off card A makes the card appear magnified when viewed from above. The magnification may be interpreted as if card A is located above its actual location. Cards B and D are mounted so that half of the cards are above the waterline. When viewed from the top of the tank, the bottom halves of the cards are not visible due to total internal reflection. Cards B and D, however, are entirely visible when viewed through two parallel sides of the aquarium. The portions of the cards B and D below the waterline are not visible when viewed through two sides of the vessel that are at a right angle with respect to one another. This is due to total internal reflection. Having cards B and D positioned at right angles to one another allows one to compare the bottom portion of both cards simultaneously. The top portions of cards B and D are visible from all angles except when viewed from below the waterline. Cards C and E are both located below the waterline, however, card E is visible from all angles. In contrast, card C is not visible when viewed from above the waterline. Card E is coated with petroleum jelly, which eliminates the card–air–glass interface that causes the total internal reflection in the other cards. The index of refraction of the petroleum jelly is nearly the same as the index of refraction of the glass. Consequently, the light reflected off card E does not exceed the critical angle whatsoever.

Materials | Aquarium or large square battery jar, 5- or 10-gal

Permanent marker

Petroleum jelly

Playing cards, 5 (face cards are preferred due to detail)

Tape, clear packing

Safety Precautions | Although this demonstration does not pose any obvious risks, always follow laboratory safety rules.

Preparation | Use a marker to indicate the water fill line—marked half way up the side of the aquarium. Use clear packing tape to attach four playing cards to the outer surface of a glass aquarium with the cards facing into the tank. Card A should be taped to the bottom of the tank, facing upward; card B should be on the side of the tank so that half protrudes above the waterline; card C should be affixed to the back of the tank below the waterline; and card D should be affixed to the back of the tank so half protrudes above the waterline. Card D is rotated—just for fun. Smear petroleum jelly over the front surface of card E and affix it to the back of the tank below the waterline, being sure that all of the air bubbles between the card and tank have been squeezed out. Finally, fill the tank half full of water.

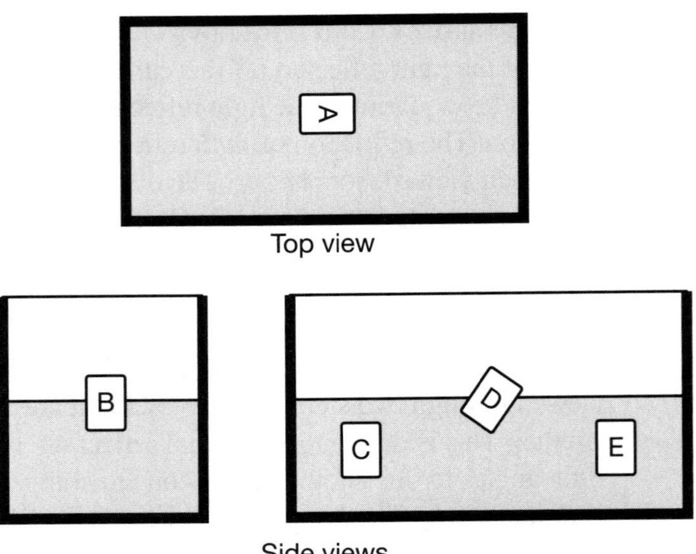

Demonstration Leave the card tank on a table in the classroom and allow students to explore the device over the course of a day or two. Upon returning to class, ask students to describe their observations of the tank. Discuss the reflection and refraction of the cards.

Reference Jones, Brian. *Little Shops of Physics;* Colorado State University: Fort Collins, CO.

A convex lens is used to project a real image onto a surface.

Application | Real image • Convex lens • Focal length • Aberration apertures

Theory | A convex lens may be used to produce a real image of an object that is inverted. Arrows are placed on the surface of the lightbulb used in this demonstration to allow students to visualize the inversion of the image. A real image is one that can be focused onto a surface. Unlike a virtual image, the light rays of a real image intersect in an organized fashion on the surface, and a real image can be projected onto a surface. The position of the image with respect to the object may be determined using the lens equation: $1/f = 1/d_i + 1/d_o$ where d_i is the distance between the image and the lens, d_o is the distance between the object and the lens and f measures the focal length of the lens. The focal length is proportional to the curvature of the lens and dictates the degree of magnification of the image. The maximum size of the image occurs when the object is located near (but not at) "F," the focal point. As the object moves away from f, the image size decreases. At $2f$, $d_i = d_o$. If the object is placed between f and the lens, a real image is not produced. The lens equation shows that when $d_o < f$, the image distance is negative, indicating that the image is virtual. The virtual image may be viewed by looking through the lens toward the object.

When the object in front of the lens is raised, the light rays are shifted, resulting in the image moving downward. When the lens is moved upwards, the image moves with it. Both of these changes may be illustrated using ray diagrams.

When a mask covers half of the lens, only half of the light reaches the lens, yet the full image is still visible, although it is dimmer. The image remains intact because light is coming from each part of the object to each part of the lens. Covering part of the lens does not eliminate any single part of the object—it only decreases the total amount of light. However, if the mask is placed between the lens and the screen, part of the image will be blocked. The lens serves to collimate the light from the object into a specific pathway. Blocking part of the pathway will block part of the image. The quality of the image cast by a lens depends upon the workmanship of the lens. An inferior lens will produce a poorly focused image known as an aberration. Adjusting the aperture in front of the lens improves the focus by limiting the amount of light received and emitted by the lens. Decreasing the aperture eliminates some of the light that contributes to the image appearing out of focus. Although the image may now appear more in focus, it is also dimmer, showing that the improvement is not without consequence. The apertures of optical instruments are measured in terms of "f-stop numbers." The f-stop number is the ratio between the focal length of the mirror and the diameter of the aperture.

Materials

Drawing compass

Lamp socket with switch and cord

Lightbulb, tubular, 25-watt

Magnifying glass, 4–6″ diameter

Pencil

Poster boards, 8″ × 10″, 2

Right angle clamp

Ring stands, large, 2

Tape, electrical

Utility clamp

Utility knife

Preparation

Use a compass to outline a 1″ hole in the center of an 8″ × 10″ piece of poster board. Use a utility knife to cut out the opening. Cut two thin arrows from electrical tape and mount one lengthwise along the top surface of the bulb. Mount the second arrow perpendicular to the first. Mount the magnifying glass onto a ring stand using a right angle clamp. Mount the lamp socket to a ring stand using a utility clamp. Position the lens between the lamp and projector screen (or wall). The lens and lamp should be aligned along a common principal axis. Prepare for the first part of the demonstration by adjusting the size of the image to 2–3X in magnitude.

Demonstration

Turn on the light source and ask students what they see. Discuss how the lens inverts the image and the fact that the image appears larger than the object. Ask students what will happen if the light source is moved closer to the lens. Following discussion, move the light source towards the lens and observe that the image now appears even larger, but it is out of focus. Ask students what can be done to refocus the image. Discuss the lens equation and explore the different combinations of object and image distances to achieve a focused image. Place the object between the lens and its focal point to show that a real image is no longer projected. The lens equation will show that an image occurs at a negative distance, indicating that the image is virtual.

Ask students what will happen to the image if the object (light source) is moved up or down. As the object moves upward, the image will move downward. Now explore what will happen if the lens is moved in a similar fashion.

Ask students what will happen if half the lens is covered with a section of poster board placed in between the object and the lens. After testing their hypotheses, discuss how each part of the lens collects light from the object and forwards it toward the screen. Now insert the poster board into the path of the light rays emitted from the lens to show that the rays can be blocked on the image side of the lens.

Ask students what will happen if the aperture is placed between the object and the lens. Insert the aperture to show how the image is improved by reducing some of the diffused rays that distort the image.

Dim the lights and use the magnifying glass to project the image of a window scene or doorway onto a screen or wall. The demonstration is most effective when there is a large contrast in light between the room and the light source (window/door). Direct students' attention to the details that are visible in this image, including the color and movement of objects past the window or door.

Chapter 20

Interference and Diffraction

Wave Theory of Light

A double slit is used to demonstrate that light behaves like a wave.

Application | Wave theory of light • Interference

Theory | If a series of particles is projected toward two closely spaced slits, then the resulting pattern on a screen behind the slits will consist of two intact lines. In 1801, Thomas Young discovered that when monochromatic light passed through two closely spaced slits, the pattern produced suggested that light is a wave. Unlike the pattern produced by the particles, the light that passes through two slits produces a series of lines—not just two. In this demonstration two concentric patterns on overhead transparencies are laid on top of one another to illustrate how light waves interfere constructively and destructively with each other.

Materials | Crossbar

Laser pointer, green

Ocular lens from a microscope

Overhead projector

Overhead transparencies of the concentric patterns, 2

Right angle clamp

Ring stands, 2

Slide with a double slit, 35-mm (e.g., slit width = 0.1 mm; spacing = 0.3 to 1.5 mm)

Utility clamps, 3

Safety Precautions | Never look directly into the beam of a laser or direct the beam toward anyone's eyes. Do not use lasers that exceed 5 mW in power. Always follow laboratory safety rules whenever performing demonstrations.

Preparation | Mount the laser and ocular lens on a crossbar using two utility clamps. Adjust the laser so its beam will pass through the lens causing it to widen. Mount the slide with the double slit to a second ring stand. Aim the laser beam through the double slit and project the interference pattern onto a white background or screen. The farther the slit is from the screen, the wider the pattern. Prepare two overhead transparencies of the concentric patterns (page 349).

Place one concentric pattern on the overhead and project its image on a screen. Discuss how the projection serves as a model of a light wave emanating from a single slit. Place the second pattern onto the first so the slits are not on top of one another. Direct students' attention to the interference pattern projected on the screen. In particular, point out the areas of constructive and destructive interference. Vary the distance between the slits and observe how the pattern changes. Remind students that this projection is simply a model and then ask them to imagine what they think would be observed if the rays of light were bisected by a screen. Use the laser apparatus to verify students' predictions. Turn on the laser pointer and observe the interference pattern projected on the wall. The pattern will consist of alternating bands of light and dark spaces.

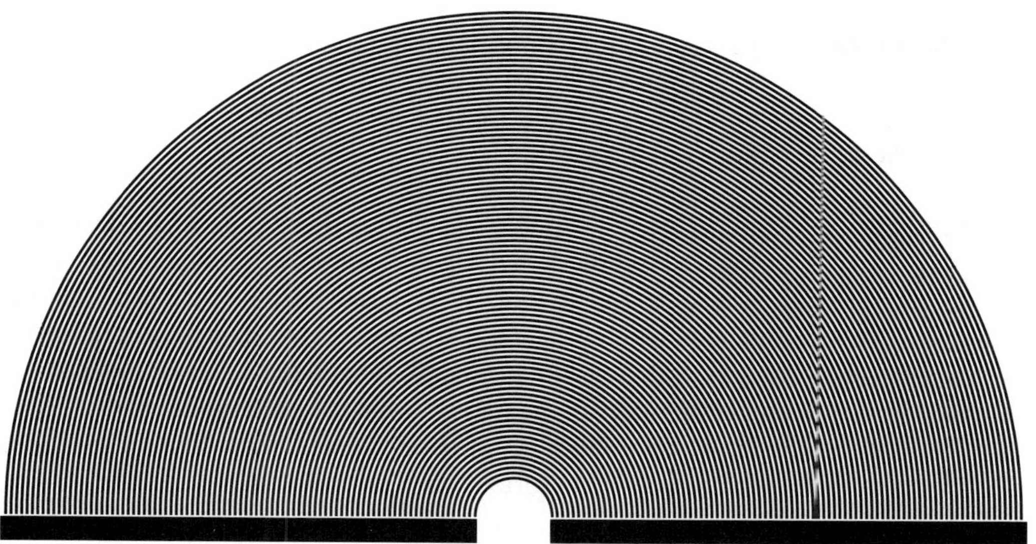

Two-Finger Diffraction

A diffraction pattern is observed using two adjacent fingers.

Application

Single slit diffraction • Wave theory of light

Theory

The divergence of light from its initial direction of travel is called diffraction. Diffraction occurs when light waves travel through small openings, around obstacles or around sharp edges (e.g., a razor blade). The bending of light causes it to interfere with itself constructively and destructively, producing a series of bright lines or dots called maxima with dark bands or regions in between. The appearance of the diffraction pattern reinforces the wave nature of light. In this demonstration, light passes through the narrow opening between two adjacent fingers, producing this pattern.

Materials

Light source (e.g., fluorescent lightbulb of a desk lamp)

Safety Precautions

Although this demonstration does not pose any obvious risks, always follow laboratory safety rules.

Demonstration

Turn on the light source and then darken the room. Ask students to hold their forefinger and middle finger together and bring them about 5–10 cm from one eye. Instruct students to look through the narrow space between the first and second knuckles. Students should see a pattern of dark and light lines in between their fingers with a pronounced dark line in the center of this pattern. Instruct students to adjust the spacing in order to bring the pattern into view. The space should be about the thickness of a sheet of paper. Discuss how light bends around the edges of an opening to yield a diffraction pattern.

Poisson Spot — The Inner Light

A laser beam diffracts around a penny revealing a Poisson Spot.

Application | Poisson Spot • Diffraction • Wave theory of light

Theory | The divergence of light from its initial direction of travel is called diffraction. Diffraction occurs when light waves travel through small openings or around obstacles and sharp edges, such as a razor blade. In this demonstration the widened beam of a laser is focused on a penny, producing a shadow of the penny. A diffraction pattern of concentric circles will appear inside the shadow, with a pronounced spot of light, called the Poisson Spot, located in the center of the pattern. The appearance of the spot reinforces the wave nature of light.

Materials | Adhesive, construction

Bamboo skewer

Cross bar

Laser pointer, green

Modeling clay

Ocular lens from a microscope

Penny

Right angle clamp

Ring stand

Slotted mass, 100 g (or similar object)

Utility clamps, 2

Safety Precautions | Although this demonstration does not pose any obvious risks, always follow laboratory safety rules. Laser light can cause permanent damage to eyes. Do not direct a laser beam toward anyone's eyes. Do not use lasers that exceed 5 mW in power.

Preparation | Mount the laser and ocular lens on a crossbar using two utility clamps. Adjust the laser so its beam will pass through the lens causing it to widen. Mount the penny to a bamboo skewer using construction adhesive. Mount the skewer to a slotted mass as illustrated in the diagram. Aim the laser beam toward the penny. Adjust the width of the beam so it is wider than the penny. Project the shadow of the round penny onto a white surface or screen. The farther the penny is from the screen, the wider the shadow.

Demonstration | Invite students to stand near the projection screen for a best view. Conduct this demonstration in a dark room. Turn on the laser beam and adjust the distance between the penny and lens until the Poisson Spot is most pronounced. Ask students if they can explain why a spot of light is visible in the center of the shadow of the penny. Discuss how the light diffracts around the edges of the penny.

Bragging about Diffraction

A diffraction grating is used to determine the wavelength of a laser beam.

Application | Diffraction • Bragg equation • Wavelength

Theory | A diffraction grating consists of a piece of plastic (or other transparent material) with thin parallel scratches spaced out at regular intervals like the lines in a notebook. Light that passes through each of these thin slits diffracts or bends around the corner in a predictable way. The bending of light causes it to interfere with itself constructively and destructively, producing a series of bright lines or dots called maxima with dark bands or regions in between. Constructive interference creates the maxima whereas destructive interference results in the dark regions. The brightest maxima, called the Zero Order Maxima or $m = 0$, is in line and opposite the slit. Other maxima are located at an angle with respect to the original path of the incoming light.

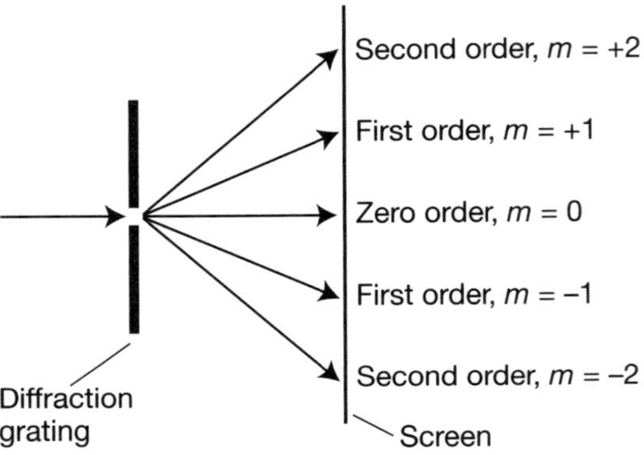

The amount of diffraction, measured as θ, depends upon the width of the slit, d and the wavelength of the incoming light, λ:

$$d(\sin\theta) = m\lambda$$

In this case, m refers to the order number. The order number is matched with θ. That is, θ_1 produces $m = 1$, θ_2 produces $m = 2$, etc. The brightness of the maxima decreases as the order of maxima increases.

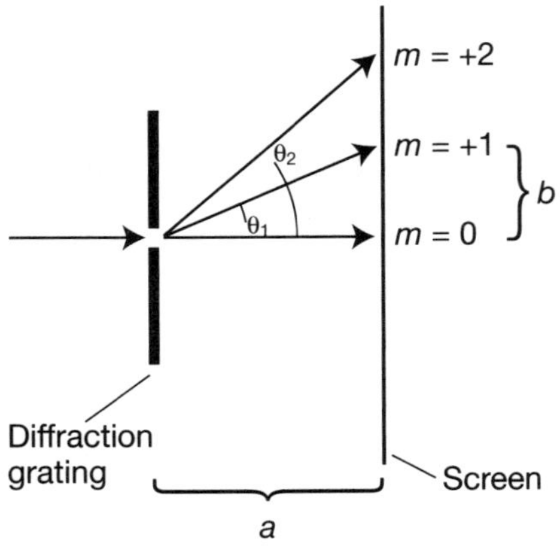

The angles of θ can be determined by measuring the distance between the diffraction grating and the screen, *a,* and the distance between the diffraction maxima (lines or dots), *b*—measured with respect to the zero order maxima, m = 0. θ can be determined using the equation:

$$\tan \theta = \frac{b}{a}$$

Materials

Crossbar

Diffraction grating, with known spacing (e.g., 600 lines/mm)

Laser pointer, labeled with wavelength

Meter stick

Right angle clamp

Ring stand

Utility clamps, 2

Preparation

Mount the laser and a diffraction grating on a crossbar using two utility clamps. Adjust the laser so its beam will pass through the grating. Project the diffraction pattern on a white background or screen. The farther the grating is from the screen, the wider the pattern.

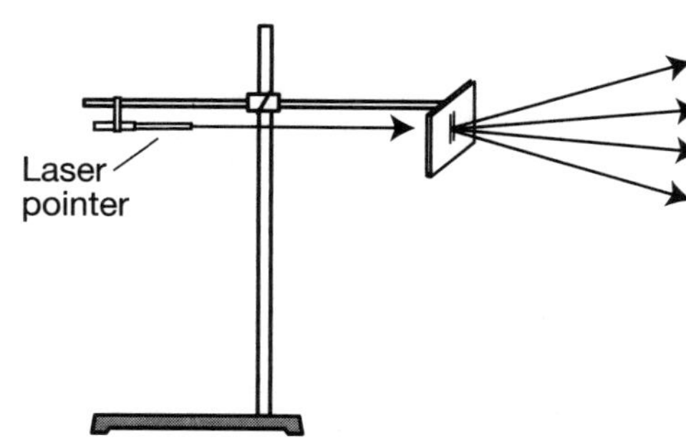

Demonstration

Turn on the laser and direct students' attention to the diffraction pattern on the white background. Move the apparatus closer and farther from the screen to show how the distance between maxima is directly proportional to the distance between the apparatus and the screen. Introduce the Bragg equation and explain how it may be used to determine the wavelength of the laser light. Use a meter stick to measure the distance between the grating and the screen, and then measure the distance between the maxima. Calculate the wavelength of the laser and compare this value to the one listed on the label of the laser pointer. As an option, the demonstration may be repeated using a different colored laser to show that the spacing between maxima is wavelength dependent.

How Thin Is the Film?

Soap bubbles are used to illustrate the optics of thin films.

Application

Thin films • Interference

Theory

In this demonstration soap bubbles are introduced into an aquarium partly filled with carbon dioxide gas. Since the CO_2 gas is denser than air, the air-filled soap bubbles will float on the CO_2. The gas is also soluble in water, however, and thus dissolves in the soap–water solution of the bubble. In turn, the CO_2 will move into the bubble cavity since the concentration of the CO_2 gas is higher on the outside of the bubble than inside. As the CO_2 enters the cavity, the bubble increases in size, causing the bubble film to become thin. Subsequently, the film changes color. The thickness, t, of the bubble may be determined using the equation:

$$t = \frac{\left(m\lambda_0 - \dfrac{\lambda_0}{2}\right)}{2n}$$

where m refers to the order (1st ring), λ_0 is the wavelength in air, and n is the index of refraction of the soap solution. In general, the wavelength decreases (towards violet) as the thickness of the bubble decreases.

Materials

Aquarium, 5- or 10-gal

Baking soda, ½ cup

Soap bubble solution and wand

Vinegar, 2 cups

Safety Precautions

Although this demonstration does not pose any obvious risks, always follow laboratory safety rules.

Demonstration

Ask students to gather around the aquarium for a better view. Tell them that you will produce carbon dioxide gas inside the aquarium using baking soda and vinegar. Pour 2 cups of vinegar into the 5- or 10-gallon aquarium. Sprinkle ¼ to ½ cup of baking soda over the vinegar to generate carbon dioxide gas. Use a bubble wand to gently blow 2–3 bubbles into the aquarium. Students will notice that the bubbles float. Discuss why the bubbles float. Following this discussion, produce a few more fresh bubbles and ask students to pay particular attention to these bubbles over time. Students may notice that the bubbles increase in size and change color as they grow. Discuss why the bubbles increase in size. Direct students' attention to the fact that the bubbles change from a greenish tint to a violet tint as they grow in size. Discuss how the wavelength of the reflected light can be used to determine the thickness of the film.

Chapter 21

Atomic and Nuclear Physics

Photoelectric Effect

Ultraviolet light is used to discharge an electroscope.

Application | Atomic theory • Quantum theory • Dual nature of light

Theory | In the photoelectric effect electrons known as "photoelectrons" are ejected by negatively charged metals when light shines on the metal. The alkali metals are particularly subject to this effect. The emission of photoelectrons depends upon the wavelength of the light cast on the metal. Red light (λ = 698–652 nm) will not cause the ejection of photoelectrons from potassium no matter how intense the light. Yet even a very weak yellow light (λ = 588 nm) shining on potassium will initiate the effect. This effect cannot be explained by classical physics, which has no quantum concept, although classical physics correctly views light as a form of energy. Einstein was able to explain the photoelectric effect by viewing light as a dual-natured phenomenon, where the energy of the light is carried by a particle called a photon. The photon energy is proportional to its frequency: $E = hf$, where h is Plank's constant (h = 6.67×10^{-34} J·s) and f is the frequency of the light. If the frequency of the light is too low, then photoelectrons will not be ejected. The frequency of the illuminating light must be greater than the minimum energy, referred to as the "work function," for photoelectrons to be emitted. Many metals require light in the ultraviolet region in order to initiate photoelectric emission. In this demonstration a glass filter that blocks UV light is placed between the light source and an electroscope to confirm that the wavelength required for discharge is in the UV range. A source of white light is also used to reinforce this notion. The electroscope is later given a positive charge to show that the photoelectric effect is also charge-dependent.

The method of discharging an electroscope can be used to confirm the unknown charge of any object. This is probably the only method of charge-confirmation that is available to most physics instructors.

Materials | Acrylic rod, ½″ diameter × 40–50 cm long (Lucite® or Plexiglas®)

Alligator clip

Banana plug

Coat hanger

Electroscope (Braun-type is recommended)

Glass plate large enough to cover UV light source (e.g., ordinary window glass)

Incandescent lamp, bright

PVC pipe, ½″ diameter × 40–50 cm long

Steel wool, fine

UV light source, short-wave (e.g., UV lamp used by geologists)

Wire cutters/crimper

Wool cloth or piece of fur

Zinc plate, 10 cm × 10 cm

Safety Precautions

Although this demonstration does not pose any obvious risks, always follow laboratory safety rules. Remind students to *never* look directly at a source of ultraviolet light.

Preparation

Polish the surface of the zinc sheet with fine steel wool. Coarse sandpaper should not be used since the rough surface of the zinc will promote quick discharge. Use a pair of wire cutters to cut a 5-cm piece of coat hanger wire. Connect a banana plug to an alligator clip using this wire and crimp to secure them together. Attach the zinc plate to the electroscope using the banana–alligator connector. This demonstration is best performed in a dry environment, since humidity will promote quick discharge of the apparatus.

Demonstration

Charge a PVC pipe negatively by rubbing it with a woolen cloth. Transfer the charge to the electroscope by drawing the pipe across the zinc plate. Ask students how to discharge the electroscope. Discuss and try the various methods suggested by students. Recharge the electroscope and shine the UV light onto the zinc plate. The electroscope should discharge immediately. Emphasize that no object came into contact with the device. Ask students to explain this phenomenon. Discuss how the UV light liberates photoelectrons from the surface of the metal and thereby discharges the electroscope. Charge the electroscope positively using a combination of the acrylic rod and woolen cloth. Shine the UV light onto the zinc plate to show that the electroscope will not discharge because the electroscope does not have excess electrons to emit. Discharge the electroscope by touching it with your hand and then recharge the electroscope negatively using the PVC pipe. Shine a bright white light source onto the zinc plate. The electroscope will not discharge regardless of the intensity of the source of incandescent white light. Place a piece of glass over the UV light source and direct the light towards the electroscope. Direct students' attention to the fact that the device does not discharge. Remove the glass plate and observe how the electroscope discharges suddenly. This demonstrates that the photoelectric effect is wavelength-dependent since neither the white light nor the filtered UV light discharged the device.

Physics "Phriendship" Bracelets

Phosphorescence and fluorescence are differentiated from one another.

Application Phosphorescence • Fluorescence • Quantum theory

Theory Fluorescence is a phenomenon that occurs when atoms in a material absorb UV rays and emit light in the visible region of the spectrum. Fluorescence is observed in a variety of materials including certain minerals, gems, crude oil, and numerous organic compounds. In all cases, the presence of UV light is required for fluorescence to occur. Phosphorescence is related to fluorescence in that UV rays are absorbed and visible light is emitted. In contrast to fluorescence, phosphorescence is a delayed process—the UV light is absorbed and then continues to emit visible light well after the original light source has been removed. Zinc sulfide and strontium aluminate are two compounds commonly used as phosphorescent pigments in glow-in-the-dark materials and objects. It is worth mentioning here that chemiluminescence (glowsticks) and bioluminescence (fireflies) are not examples of either effect.

Materials Craft beads, UV-sensitive (available at major craft stores)

Lanyard, glow-in-the-dark, 30 cm per student (available at major craft stores)

UV light source

Safety Precautions Although this demonstration does not pose any obvious risks, always follow laboratory safety rules. Remind students to *never* look directly at a source of ultraviolet light.

Demonstration At the end of class, distribute one bead and a 30-cm piece of lanyard per person. Tell students to make a bracelet from the bead and lanyard and to wear it until the following class. Students may be told that this bracelet represents a special bond between them and physics. Play it up!

Upon returning to class, ask students to describe their observations. Some will say (incorrectly) that the bead "glows," others will say that the lanyard glows. Discuss the difference between the color change of the bead and the phosphorescence of the lanyard. Note that the beads undergo a UV-induced color change but they are not fluorescent.

A peashooter is used to simulate the Rutherford scattering experiment.

Application | Atomic theory • Nucleus of atom

Theory | In 1909, Ernest Rutherford directed an experiment known as the Geiger–Marsden experiment, which is commonly referred to as the Gold–Foil experiment. In this experiment alpha particles were fired toward a thin piece of gold foil. Most of the alpha particles were deflected at small angles upon penetrating the foil. To Rutherford's surprise, however, a significant number of particles were deflected by angles much larger than 90°—with some returning straight back. These results were completely unanticipated, upon which Rutherford later commented, "It was almost as incredible as if you fired a 15-inch shell at a piece of tissue paper and it came back and hit you." Rutherford concluded that the positive charge of the atom was located in a very small region in the center of the atom.

Materials | Adhesive, construction

Dowel, $1''$, $3\frac{7}{8}''$ long

Peas, dry

Printer paper box top ($\approx 11'' \times 17'' \times 3''$)

Straw, large (to be used as a peashooter)

Tape

Tissue paper, $\approx 50\text{ cm} \times 100\text{ cm}$

Tuna fish can, empty, clean

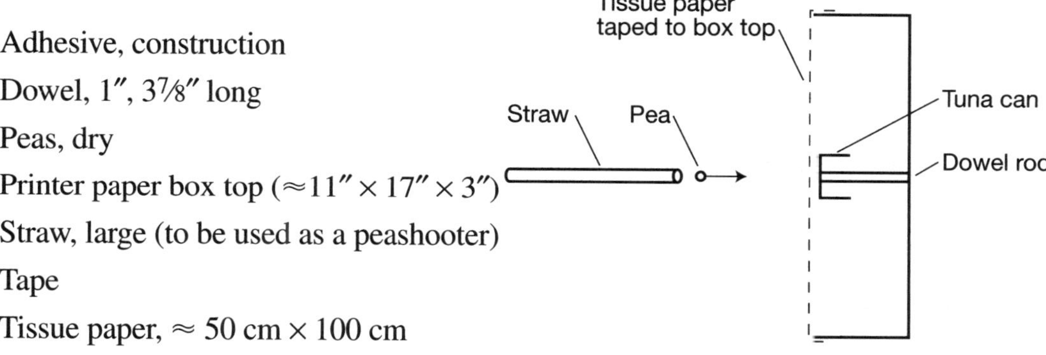

Preparation | To create a Rutherford simulator, use construction adhesive to glue a $1''$ dowel to the inside center of the top of a printer paper box. Glue an inverted tuna can onto the protruding dowel. Allow the adhesive to set overnight. Cover the opening of the box with tissue paper drawn tightly over the edges and secured with tape. The paper should barely come in contact with the bottom of the can without exposing an outline of the can. If the can is too far away, a longer dowel must be used. A box made from $\frac{1}{4}''$ plywood or Luan® may be used in place of the paper box top for a long-lasting apparatus.

Demonstration | Review the development of the atomic theory beginning with Democritus and then emphasizing Thomson's "Plum Pudding Model," which stated that an atom is a positively charged sphere with negative particles imbedded in it. Describe how Ernest Rutherford fired alpha particles at a thin sheet of gold foil and determined that the positive charge of the atom resided in a very small region in the center of the atom. Prop the Rutherford simulator against a wall or ring stand. Simulate Rutherford's experiment by using a peashooter to launch peas toward the tissue paper–covered box. Most peas will disappear behind the tissue paper while those that strike the can will bounce back toward the shooter.

Blackbody Radiation

A homemade device is used to illustrate blackbody radiation.

Application | Blackbody radiation • Wien's Law • Pigment

Theory | A blackbody is an object that absorbs all radiation that falls upon it. In order for an object to be seen (visible), light must reflect off of it. For example, an apple is seen as a red apple under white light because red light reflects off its surface while the blue and green light waves are absorbed by the pigment in the apple. An object such as black paper or ink appears black because the pigment absorbs nearly all of the white light. Most objects that are said to be "black" are not truly black but actually nearly black since they still reflect enough light to be visible. Only those objects that do not reflect light are true blackbodies. Just like glowing, red-hot coal, blackbodies are seen by the light they emit, produced by their heat. According to Wien's Law, the temperature, T, of an object is directly related to the color of light emitted by the object.

$$\lambda = \frac{2.90 \times 10^{-3} \text{ m·K}}{T}$$

For example, the light emitted by the Sun is in the range of 500 nm, indicating that the temperature of the surface of the Sun is approximately 6000 K.

Materials | Adhesive, construction

Compass

Frame matting, 8″ × 10″, 4 pieces
- Dark gray matte finish (almost black)
- Light black matte finish
- Darkest black matte finish
- Pure white matte finish

Frame matting, pure white, matte-finish, 9″ × 42″

Printer paper box

Ruler, cm

Utility knife

Safety Precautions | Although this demonstration does not pose any obvious risks, always follow laboratory safety rules.

Preparation | Use construction adhesive to secure an 8″ × 10″ pure white, matte-finished matting onto the side face of a printer paper box. Insert a 9″ × 42″ piece of pure white, matte-finished matting inside the box as illustrated in Figure 1. Secure with construction adhesive and allow to set overnight.

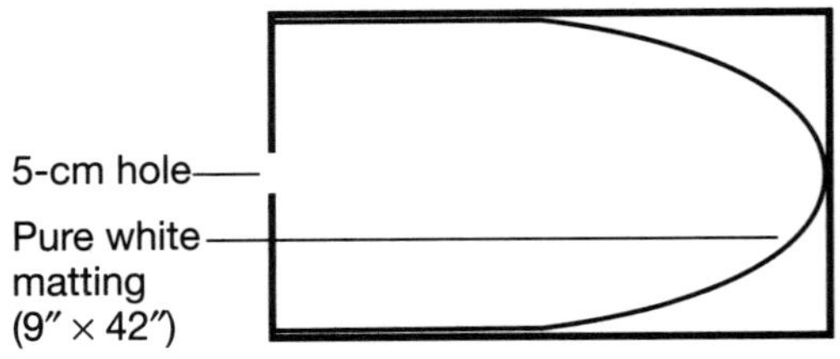

Figure 1. Top view.

Cut a 5 cm hole into the center of the white matting and box face. Cut a ⅝″ × 8½″ wide strip in the lid of the box near the edge above the hole. Replace the lid on the box with the strip near the hole. Insert the four 8″ × 10″ mattings through the slot in the following order: white, gray, light black, and dark black, with dark black being the innermost (see Figure 2).

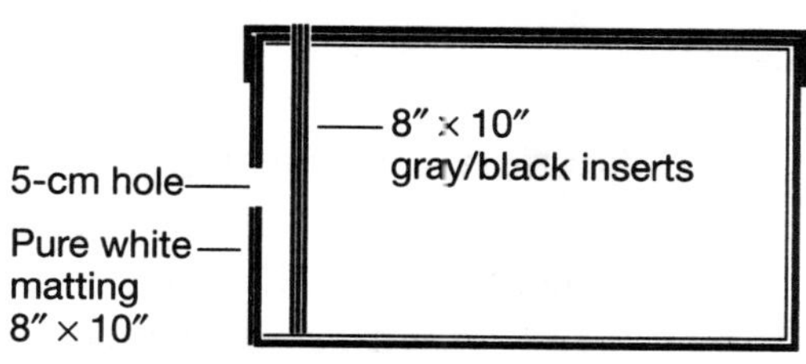

Figure 2. Side view.

Demonstration

Place the box at least 2 meters away from the audience. Direct students' attention to the hole in the box and ask them what color they observe. Remove one matting at a time from the slot and each time ask students to describe the color of the hole. Ask students to identify the darkest hole. It is expected that students will identify the hole without the matting as the darkest of all. At this point, remove the lid from the box and reveal the fact that the inside of the box is actually white. Discuss how the darkest hole observed demonstrates a complete absence of light.

Emission Spectra

The Balmer series of the hydrogen atom is compared to a continuous spectrum.

Application	Atomic theory • Emission spectra • Bohr model • Quantum theory
Theory	In 1913, Niels Bohr improved on Rutherford's model of the atom by stating that electrons are restricted to specific levels around the nucleus. Furthermore, electrons may move from one level to another but do so only by emitting or absorbing specific packets of energy called quanta. When an electron absorbs energy, it becomes excited and moves to a higher energy level. For an electron to move to a lower level, it must release energy. The energy is released in the form of a photon having a specific frequency of light (visible and non-visible). The visible part of the emission spectrum is called the Balmer series. The hydrogen atom emits four distinct frequencies in the Balmer series: one red line, one blue line, and two in the violet region. Bohr's discovery serves as one of the foundations of quantum theory and the modern model of the atom.
Materials	Bulb socket and cord Diffraction glasses, one per student (e.g., "Rainbow glasses") Gas-discharge tube, hydrogen Gas-discharge tube power supply Incandescent bulb, 25-watt aquarium (tube type, clear) Ring stand Utility clamp
Safety Precautions	Although this demonstration does not pose any obvious risks, always follow laboratory safety rules.
Preparation	Position the gas discharge tube and power supply vertically in the middle of the demonstration table in front of the audience. Insert the aquarium bulb into the light socket and mount the socket in a vertical position on a ring stand using a utility clamp. Position the lightbulb above and directly in line with the gas discharge tube.
Demonstration	Distribute one pair of diffraction glasses to each student. Turn on the aquarium bulb and then darken the room. Ask students to find the rainbow of light emitted by the bulb as they look through the diffraction glasses. Once all students see the rainbow, turn on the hydrogen gas discharge tube and ask students to look for the spectrum produced by the hydrogen tube located directly below the continuous spectrum given off by the aquarium lamp. Have students compare the lines of the hydrogen spectrum to the full spectrum of the white light of the aquarium bulb. Direct students' attention to the fact that the lines of the hydrogen spectrum are in line with the same colors of the full spectrum. Turn off the white light and have students observe the four visible distinct lines of the Balmer series in the hydrogen spectrum—one red line, one blue line, and two in the violet region. Discuss how the Balmer series confirms the Bohr model of the atom.

 —————————————————————————— A DEMO A DAY

The Meaning of (Half) Life

A collection of modified dice is used to simulate the concept of nuclear decay.

Application Nuclear decay • Half-life • Radiation

Theory Half-life refers to the amount of time required for an isotope to decay from its original amount. For example, the half-life of radon-222 is 3.8 days, which means that only half of the original amount of this isotope will remain after 3.8 days. Rn-222 emits alpha particles and decays into Po-218. This demonstration uses a collection of modified dice to simulate the concept of nuclear decay. Each die has a hole bored through it and when rolled it has a 1/3 chance of landing hole-side up. Each "roll" represents a definite time period. When using 64 dice the decay sequence (rounded to the nearest whole number) will be 64, 48, 32, 21, 14, etc. Half-life is reached after the second roll.

Materials Cup, 16-oz

Dice, 64, 48 or 32* (available in packages of 10 at most dollar stores)

Drill

Drill bit, ceramic, $\frac{1}{8}''$

Overhead projector and screen

Photo box frame, plastic, $8'' \times 10''$

Vise

*This demonstration should be performed with at least 32 dice, but a larger number will provide more persuasive results.

Safety Precautions Although this demonstration does not pose any obvious risks, always follow laboratory safety rules. Follow manufacturers' directions when using power tools.

Preparation Clamp one die in a vise with the single pip (pit) facing upward and drill a $\frac{1}{8}''$ hole clear through the center face of the die. Repeat with each die.

Demonstration Discuss what is meant by the half-life of an isotope. Use the dice to reinforce the concept by spilling the 48 dice into a plastic photo box frame placed onto an overhead projector. Students will notice that approximately a third of the dice will cast a shadow with a white dot in its center—representing the isotope that has decayed. Sketch the axis of a graph with the number of dice on the y-axis and roll number on the x-axis. Plot a data point representing the first roll on this graph. Remove the "decayed dice" and repeat the roll on the overhead with the remaining dice. Once again, approximately one-third of the dice will have decayed and should be removed and a second data point is added to the graph. This process is repeated until all the dice have decayed. Direct students' attention to the exponential decay curve of the data.

Reference Richard Berg, Department of Physics, University of Maryland, Baltimore.

Mass Curves Space

A homemade apparatus is used to illustrate how mass curves space.

Application	Space–time distortion • Curvature of space • Cosmology
Theory	Albert Einstein's Theory of General Relativity predicts, among other things, that space curves in the presence of mass. The amount of curvature is directly related to the size of the mass. This demonstration uses a sheet of cloth upon which a reference grid has been drawn. The sheet represents a cross section of space. Placing a large mass onto the cloth distorts the grid pattern causing the pattern to curve around the mass. The distortion is similar to the effect described by Einstein. According to Einstein, gravity is a consequence of this phenomenon. An object traveling near a massive object in space will be drawn toward it, causing its path to curve around the massive object. Einstein's idea, however, does not contradict Newton. Newton explains that the gravitational pull between the two objects causes the smaller object to move in a curve around the larger object. That is, according to Newton, the object leaves its straight-line path because an outside force acts upon the moving object. In comparison, Einstein says that the object continues to travel along the straight line, but space–time itself is being curved which causes the smaller object to move around the larger mass. If the smaller object is close enough to the larger mass, then the object can go into orbit around the larger mass or actually fall inward.
Materials	Black marker Embroidery hoop, 12″ or larger Marble, ½″ wooden or glass Ring stands and rings, 3 Ruler Spandex® cloth, thin, light-colored, 16″ × 16″ piece Steel ball, 1–1½″
Safety Precautions	Although this demonstration does not pose any obvious risks, always follow laboratory safety rules.
Preparation	Use a permanent marker and ruler to draw a 2-cm × 2-cm grid across the surface of a sheet of Spandex. Stretch the Spandex across the embroidery hoop and secure it with the outer hoop. Rest the hoop horizontally and level onto three rings mounted to stands. Place the steel ball into the center of the Spandex to verify that the cloth stretches adequately. Placing the ball on the cloth should form a vortex shape. Adjust the tension of the cloth, if necessary. Remove the ball prior to the demonstration.

Roll a marble along a grid line across the surface of the Spandex and discuss how Newton's First Law supports the observation that the marble traveled in a straight line. Discuss Einstein's proposal that mass curves space–time. Place a large steel ball in the center of the Spandex and observe how the mass curves the grid. Have the students observe the curve from the side also. Ask students what will happen if the marble is aimed along the straight portion of one of the grid lines. Roll the marble along the straight portion of a grid line and observe how it then follows the curved portion of the grid line as Einstein predicted. If the marble is aimed close to the mass, then it will begin to orbit the mass until it falls into the vortex.

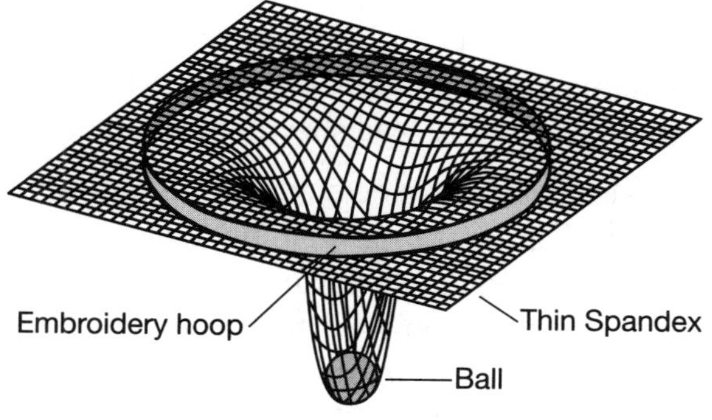

Emit It!

Various samples of radioactive objects are tested and their radiation type identified.

Application

Alpha radiation • Beta radiation • Gamma radiation

Theory

There are three types of ionizing radiation—alpha particles, beta particles, and gamma rays. Alpha particles are identical to a helium nucleus and are the most massive of the three. A beta particle is either an electron or positron. A positron has the same mass as an electron but it is positively charged. A gamma ray is a photon which is a mass-less particle. These three types of radiation penetrate matter to different degrees. Alpha particles cannot penetrate paper because they are too large. The smaller beta particles penetrate paper, but they may be blocked by the crystal structure of aluminum. Gamma rays, consisting of mass-less photons, can be blocked by only lead sheeting and other materials with a similar or greater density.

Household smoke detectors are an example of an alpha emitter or source. They contain an ionization chamber that operates using the alpha source, americium-241. As smoke enters the chamber, it becomes ionized by Am-241, causing the smoke particles to become electrically conductive and thus closing a circuit that sets off the alarm. A beta source can be found in ice-melting salt. Ace Hardware's Pet Friendly Ice Melt consists of 70% KCl. Naturally, KCl contains 0.0118% of the radioactive isotope postassium-40, which is a beta emitter. Orange–red colored Fiesta dinner-ware plates first introduced in 1936 obtained their orange color from the uranium oxide pigment used in their glazing. Uranium oxide is a gamma emitter.

Materials

Alpha source (e.g., smoke detector)

Aluminum flashing, $\approx$ 5 cm $\times$ 5 cm, 5 pieces

Beta source (e.g., Morton Lite-Salt™ or Ace Hardware's Pet Friendly Ice Melt)

Gamma source (e.g., Orange–red Fiesta™ old dinnerware plate or old glow-in-the-dark watch)

Geiger counter

Index cards, 5

Lead sheet, 5-cm $\times$ 5-cm, 2 pieces

Commercial samples of alpha, beta, and gamma sources

Safety Precautions

Although this demonstration does not pose any obvious risks, always follow laboratory safety rules. Never use radiation sources that can leave residues on hands or surfaces. Use sources of radiation that do not exceed emissions of 1 μC. Do not ingest sources of radiation.

Exhibit the three sources of radiation: alpha, beta and gamma. Show students that each sample will excite a Geiger counter. Place up to five index cards over each source and show that the index cards limit the emission of only the alpha source. Discuss why the alpha particle is blocked by the paper. Remove the index cards and place up to five pieces of aluminum flashing over each source and show how only the gamma penetrates the aluminum sheeting. Cover the gamma source with one or two lead sheets to show how lead can be used to limit gamma radiation. Finally, use these techniques to identify the type of radiation emitted by each of the common household samples.

Notes

A

Aberration
    A Real Lens 345
    It Is Real! 329

AC current
    Build a Simple AC Motor 313

AC vs DC current
    AC Blinkies 284

Acceleration
    Change Your Slope 20
    Inertia—Keeps Things Put! 56
    Newton's Jerk 73
    Work is Going My Way 78

Acceleration due to gravity
    Ball in the Cup 157
    Falling Down 29
    Falling Together 27
    Floating Down 31
    Galileo's Waterfall 35
    Heavier—Not Faster 33
    Money from the Sky 34
    A Shapely Planetoid 142

Accuracy
    How Close Can You Get? 12

Action-reaction force pairs
    Just Stepping Off 67

Adiabatic expansion and compression
    Pressing the Temperature 235
    Refrigerants Are Cool Gases 233
    Air Capacitor 254

Air resistance
    Falling Together 27
    Floating Down 31

Alpha radiation
    Emit It! 366

Alternating current
    AC Blinkies 284
    Build a Simple AC Motor 313

Altitude measurement
    How High? 2

Amplitude
    Seeing Sound 212
    Singing Rods 219
    Types of Waves 204

Amusement park rides
    Loop the Loop 140

What Makes the World Go Around 136

Angular momentum
    Spin 'em Up 166

Angular velocity
    Spin 'em Up 166
    Spin-offs 134
    What Makes the World Go Around 136

Apertures
    A Real Lens 345
    Are We Adding or Subtracting? 318
    It Is Real! 329

Astronomy
    How Wide? 3
    Atmosphere Keeps Me Down 186

Atmospheric pressure
    Atmosphere Keeps Me Down 186

Atomic Theory
    Blackbody Radiation 360
    Emission Spectra 362
    Photoelectric Effect 356
    Physics "Phriendship" Bracelets 358
    Target Shoots Back 359

B

Balmer series
    Emission Spectra 362

Baseball, physics of
    Curve Baller 196
    Standing Waves of a Baseball Bat 214

Battery
    Soda Bottle Potential 271

Beats
    You Can't Beat a Singing Rod 221

Bernoulli effect
    Curve Baller 196

Beta radiation
    Emit It! 366

Bioluminescence
    Physics "Phriendship" Bracelets 358

Bohr Model of Atom
    Emission Spectra 362

Bragg equation
    Bragging about Diffraction 352

Buoyancy
    Balls in Beans 192
    Fingering Out Buoyancy 188
    Giving Soda Pop a Lift—A Twist on an Old Idea 190

Topic Index

D

Topic Index

Topic Index

Topic Index

Topic Index

Q

R

Topic Index

Topic Index